Podcasting for Beginners

How to Start and Grow a Successful and Profitable Podcast

By
Daniel Hunt

circumstances is the author responsible for any losses, direct or indirect, that are incurred as a result of the use of information contained within this document, including, but not limited to, errors, omissions, or inaccuracies.

Table of Contents

Introduction

The advent of the digital economy has resulted in a number of new avenues opening when it comes to money-making options. Indeed, the very nature of media and marketing has changed. A field such as marketing, which was the preserve of psychologists and artistic types, has now become an intensely analytical and scientific pursuit.

These days, if you want to make money online, there is no end to your array of options. You can start a blog, start a YouTube channel, promote products on social media, advertise on Google, run paid ads on Facebook and other platforms, host webinars, run affiliate marketing, build niche websites... I mean, the choices are endless. In the midst of all of these fantastic choices, there remains one that is often overlooked: Podcasting.

Almost everyone has heard of it, but despite this, podcasting remains a curiously underserved field. There are quite a number of podcasts up and running, and it isn't as if there's no competition, but you'll be hard-pressed to find a wannabe entrepreneur lining up to launch her own podcast. Why is this?

Well, if you ask me, it's a combination of two things. The first is that there is a technical aspect to launching a podcast that acts as a screener. Unlike writing a blog, one simply cannot jump in and start recording themselves and call themselves a host of a podcast. There's a bit more work that goes into it. You need to have a solid background in terms of producing the show, mastering the audio, and uploading your files correctly.

Of course, all of this comes before you even begin marketing your podcast. Marketing a blog that has an inherent web presence is tough enough, but how does one begin to market an audio show? People don't have the patience to watch seven-minute videos these days, so how are you going to get them to listen to a show that lasts for a half hour or longer?

Well, this is where I come in. To be more precise, this is where this book is going to help you. The usual books on podcasting deal heavily with the planning of your podcast and help you brainstorm topics and give you a few cursory notes about marketing and the technical stuff and leave it at that. Well, that's not what I'm going to do here.

Planning the topic of your podcast is important, but to be honest, this is not the only thing you should be focusing on. You see, selecting a great topic to talk about is not the only thing you need to do. Some people think that just because they're talking about all of this great stuff and providing information, people will automatically sign up and listen to their show. This is hardly the case.

You need to take care of a number of things after choosing your topic. Most importantly, you need to figure out how you're going to monetize your show. I mean, that's what this is all about, right? What's the point of having a wonderful podcast and spending all this money getting set up if you cannot monetize it? Unless you're loaded or really don't need the money, this just isn't practical.

The technical side of things might cause you to doze off to sleep, but it is still important for you to grasp. Not enough time is spent discussing this stuff, and I feel it is a mistake. Your listeners have just one (or at the most two) forms of media to engage with you, with audio being the primary mode of engagement. Screw this up and you might as well be a mime. And no one loves mimes.

Don't let the technical stuff scare you away, though. This isn't a book for audio geeks. At its heart it is still all about teaching you how to market and make money from your podcast. There is no shortage of channels on which you can market your podcast. As you'll see shortly, this is both a good and a bad thing. If there are many channels for you to market your podcast, there are even more ways of monetizing your podcast.

Monetization is a tricky issue because you need to know when you can switch it on, and this book will help you recognize that ideal moment. Sure, it is possible for you to start monetizing right off the bat, but this isn't always a good idea, despite the positive cashflow it brings.

New to Podcasting?

I've been addressing those people who already know what podcasting is. Of course, there might be some who have simply heard of the term and are wondering if it can help their business in any way. Well, the answer is a resounding yes! The first chapter of this book is designed especially for you if you're one of these people.

Podcasting is a great way to get the word out about your business and your products. Even if you're aiming to be an influencer and not a traditional business, podcasting is a no-brainer. Traditionally, businesses try to spread the word about themselves through advertising. However, this is a model from the pre-digital revolution era.

This is the age of free traffic and inbound marketing, which refers to when your customers search for your product and come to you voluntarily. Outlets such as blogs and podcasts are invaluable for this. A podcast is a piece of online real estate you own, and it simply makes no sense for you to ignore this potential goldmine for your business.

As you read this book, you will recognize how you can leverage the power of podcasting to provide unlimited traffic for your business and take your sales to new heights. Having said all that, I must warn you that podcasting is not a get-rich-quick scheme. It isn't something you can plug in and then expect to bring instant income to your business or online ventures.

Podcasting takes work, and there is an artistic side to it. You will need to brainstorm editing procedures and other ways of providing your listeners with a good experience. You will need to have backup content channels to support your podcast. Think of it as a powerful tool in your arsenal as opposed to your only tool.

How well your podcast works for you is really up to you and the amount of work you're willing to put in. There is some stuff you need to master, and it will seem intimidating, but really, it's not rocket science. With some hard work and dedication, you'll have a successful podcast up and running in no time.

So let's dive into the history of podcasting and take a look at where and how the term came to be!

Chapter 1: So What is Podcasting?

Podcasting has actually been around for far longer than you might guess. While the term itself has been around for a little over a decade, the idea of it, or the spirit of it if you will, has been around since the first days of the internet. At its core, a podcast is the digital equivalent of old-school radio. It functions along the same lines in terms of the business model and the appeal it has for its listeners.

It's just that beneath these similarities, everything else is different.

History

Back when the internet began to take shape in the public consciousness – this would be around the mid-90s – some clever soul figured people would like to read stuff on it instead of just perusing, shall we say, extracurricular stuff involving sex. Whoever this person was, they were right, and soon blogs came to be.

These first blogs were different creatures from the monstrosities they are today. While today's blogs are publishing giants in their own right, back then a single HTML page counted as a blog, and there was no way to know if anyone was even reading your stuff. Then there was the issue of being found on the internet.

Search engine giant Yahoo soon found a way to solve this, and blogging became a nice way to spew something onto the internet and have people respond and react to what you wrote. It was hard to ignore the consumption pattern of people who read these blogs. It was mostly as a pastime and, at best, it was a distraction. A lightbulb went off in some people's heads and they figured out that a better way to engage their readers or reader, as was likely the case, would be to convert their text into speech and somehow distribute this.

This dream became a reality in 2000 when former MTV VJ Adam Curry and the programmer David Winer developed the Really Simple Syndication feed or RSS feed, which you might have encountered when reading blogs (Geoghegan & Klass, 2008). The RSS feed made it possible for people to distribute both written and audio content on the web, and thus audioblogs came to be. Things continued like this for a while, and then Steve Jobs came along and as was his wont, shook things up.

Birth of the Podcast

Jobs didn't have a direct effect on the audio blogging world, but the release of the iPod shook things up in the music and MP3 player industry enough that bloggers could not ignore the volume of people using iTunes and the iPod. Curry was again at the forefront of this revolution and was the first person to develop code (yes, he wrote it himself), which would sync audio blogs onto his iPod (Geoghegan & Klass, 2008).

Now, Curry was not a developer, and in his own words his 'ipodder' software 'sucked' (Geoghegan & Klass, 2008). However, he made it open source and this attracted a bunch of programmers who hadn't initially seen the extent of Curry's vision. As the development world went to work, the ipodder evolved into the "Daily source code" and soon, audio bloggers everywhere were able to provide their listeners a tool that would transfer these blogs onto their iPods.

The revolution was well under way by the time journalist David Slusher remarked on this phenomenon of 'podcasting' and the word went viral (before virality was even a thing) (Geoghegan & Klass, 2008). The popularity of podcasting was big enough that in 2005, Apple integrated daily source code into iTunes and enabled bloggers everywhere to upload their audio files into the iTunes store.

Some guy who was prescient enough in 2002 to buy the domain 'podcasting.com' got paid millions for his purchase, and, thus, the first podcasts were born. These were initially just migrated radio shows. For example, a regular drivetime radio show you listened to in your car would be made available on iTunes and you could listen to it later.

Sports radio was the first to pick up steam, but the rise of Sirius XM masked the growth of podcasting (Geoghegan & Klass, 2008). Thanks to the mainstream still not being aware of the huge potential of podcasting, Sirius' subscription-based radio was attracting a lot of attention, as was the fact that Howard Stern had decided to syndicate his show on that platform. As a result, podcasting flew under the radar and was being dominated by a few pioneers.

Either way, podcasting really blew up when the iPhone was released and people now had access to the internet on the same device they carried around in their pockets. No longer did you need a separate device to listen to music and radio; it was all integrated into one sleek product. This gave rise to companies like Spotify and Soundcloud, who capitalized on this opportunity and online streaming became a thing. Suddenly, traditional radio was dead and podasting's true potential came to light.

At the beginning of this decade, major media houses had no clue as to what they could achieve with podcasts. Fast forward to today and most of them still don't have a clue. The real visionaries of podcasting are those who are nimble and are able to monetize their podcasts beyond the traditional sponsorship and advertising model.

This is a holdover from the days of traditional radio, and while it works well for niche podcasts and ones with huge fanbases, starting a podcast to simply attract advertising dollars is to limit yourself. You see, a podcast must be viewed as part of a larger content dissemination strategy.

According to research, it takes five touches or contact points for a potential customer to turn into a paying one (Geoghegan & Klass, 2008). This means you need to contact them five times and give them value to convince them that you know what you're talking about. Your customer could provide you value in terms of both money (which was all they could exclusively give you in older business models) or time.

Time spent on your platform or engagement is a big deal, and it is what drives all of the major websites today. The greater the levels of engagement, the more money you will make. This is what leads to astronomical valuations of social media companies that don't make a single cent. Money isn't the bottom line on the internet, it is engagement. The fact that a company/platform can retain its users despite the myriad number of free options points to a powerful force it wields over its users (Geoghegan & Klass, 2008).

I'm mentioning all of this because it is important for you to set yourself up the right way in terms of your mindset right off the bat. It isn't enough for you to just put out a podcast that people will listen to. You need to back it up with the other four touches. This could be a blog, a YouTube channel, a social media account, and some other form of content release (say updates on Instagram or doing an AMA on Reddit).

Monetization is important, but it should not be your only focus. The history of podcasting teaches us that engagement is what drives dollars. So focus on engagement and the rest will fall into place. Focus on quality content that your listeners love and people will come. Mind you, you can promote content that provokes outrage to drive engagement as well. The rise of right-wing celebrities like Alex Jones and Tomi Lahren shows that there are enough nutjobs out there who'll believe anything they hear and engage with you. Of course, the left has its share of nutters as well, so don't think I'm being biased here.

This isn't a book on the greater good, so it's entirely up to you to figure out what your engagement strategy needs to be. Every marketing platform has its own character as you'll see in the second half of this book, and you need to tailor your show to match that. So what can podcasting do for you, exactly? Well, let's look at this now.

The Benefits of Podcasting

At first glance, if you're a small business you might think that the only way you can make podcasting work for you is by shelling out dollars to promote your company via ads on some popular show. Well, this is the uncreative way of doing this. It is certainly the easier method of popularizing your products, but it's a short-term strategy at best. You see, internet marketing has a phenomenon called inbound marketing (Geoghegan & Klass, 2008).

This term was coined by the marketing agency Hubspot to define situations where a user finds you all by themselves, and you don't have to spend any money advertising to them. In other words, they come to you. Landing on the first spot in Google's search results and someone clicking over to your website is an example of inbound marketing. In this case, the only thing that you did was to spend time creating content that provides value to your audience. The rest is done by Google for you.

Inbound marketing is not yet fully understood by businesses. The innovative ones have grasped its power and deploy a number of strategies to get people clicking onto their websites and buying their products. For example, beauty brands run active Instagram accounts, which gets their customers excited about their products and gets them buying the stuff.

Paid advertising is still powerful, but you can consider inbound marketing to be the equivalent of word of mouth advertising. In this case, search engines that have ultimate authority over who is more credible are giving you a stamp of approval. On the internet, there's no more powerful word of mouth recommendation than that. A podcast is simply a tool for inbound marketing.

It is easily distributed and is always available for your listeners to engage with. They can stop, start, pause, do whatever they want and still engage with you. A paid ad is essentially you shoving your product in front of their faces, but inbound marketing is more like you sliding into their consciousness. In case you're still skeptical, consider the following examples.

Successful Branding

"The Message" is a podcast that is focused on science fiction (Geoghegan & Klass, 2008). It is a newscast-style show that follows the journey of scientists decoding extraterrestrial messages using a variety of tools that are found in today's world. Guess who created the show? GE. That's right, the behemoth appliance, defense contractor, re-insurer General Electric.

The products that are highlighted in this show are real-world GE appliances. They don't advertise them explicitly, but every now and then, there's a timely reminder that GE's stuff is pretty cool. Another example of a successful podcast is #LIPSTORIES, which is produced by Girlboss Radio (Geoghegan & Klass, 2008).

Girlboss Radio is a major channel, but this particular show is created in association with Sephora, the cosmetics company. Sephora sponsors the show, but takes an active hand in creating the content. The premise sounds ridiculous, but it somehow works. A number of influential women talk about their lives and seek to empower other women listening to the podcast. Sephora aims to create a range of lipsticks that empower women, which sounds like some Grade A marketing bullshit, but it gets women shopping so who am I to judge?

A great example of integrating values with the theme of a company is Blue Apron's podcast "Why we eat what we eat." Blue Apron is a food delivery service that aims to deliver fresh and healthy ingredients to its customers on the go. It is a great service if you can get it in your area. The show is an educational podcast and features a number of topics that relate to food sustainability and health. Since it is created by Blue Apron, the company gets free eyeballs on its products.

There are many other companies such as ZipRecruiter, Microsoft, Slack, and so on that have their own podcasts with their own themes. It's not just huge companies but bloggers as well who use podcasts to spread the word about themselves and engage their audiences. A good example of this is Tim Ferris, who's probably too big these days to be called a blogger but started off as one.

Another popular podcaster is Joe Rogan, who switched from random game show host to a bonafide media personality thanks to his provocative stance on a number of issues. My point is that podcasting isn't something that requires a cookie-cutter approach. You can tailor it to suit your own needs and drive your message.

Consider the internet marketer Grant Cardone who uses his podcasts and YouTube channel (which contains video recordings of his podcasts) to drive investors to his real estate investment funds and gets people to sign up for his sales courses. Yup, you read that right. The sky's the limit when it comes to podcasting, and the only limit to how you can monetize your podcast is your own creativity.

So step away from the staid old model of waiting for ad dollars. Sure, it is lucrative, but understand that thanks to the power of inbound marketing, you can survive and thrive despite the lack of that sort of money.

Made for You Content

The best part about starting a successful podcast is that you don't ever need to worry about generating content if you're a small blogger or small business owner. Your audience will let you know what they want to hear and which direction you ought to take your show in. When starting off you need to have a handful of topics on hand to talk about and set up some interviews with experts in your field.

Once this is done, though, your audience will dictate your content. This feedback loop is far more open when it comes to podcasts since people think of it as being like radio. Contrast this with a blog comment. There's still a wall between you and your customer. With a podcast, you can do things like having people call in and you can answer their questions directly. All of this gets people to trust you even more and your content generates itself.

A side-effect of all this is that your listeners will begin to view you as an actual human being and not as a company. This promotes trust in your products, and you'll find that your bottom line increases as a result.

Where Can You Find Them?

Before you rush off and begin creating your own podcast, it pays to take the time to figure out what works and listen to a few good shows of your liking. In addition to preparing you for what sort of a format works best, it will also familiarize you with the various sources that cater to podcasting.

The only thing you need to listen to podcasts is an app, which enables you to download them and organizes them on your phone or computer. Here are the best apps and resources you can use to do this as of this writing.

Websites

Every podcast has a website that hosts it. If you love reading the content of a particular website, you can simply head over to it and download their podcasts. You can store the podcast as an audio file on your computer or phone and listen to it whenever you want. Of course, this is a cumbersome way of doing this and you're better off using an app if you want to listen to multiple podcasts in one place.

iTunes

The original library of podcasts is still very much alive and kicking. On the plus side, iTunes has the largest collection of podcasts thanks to the popularity of Apple's products. The negative side of this is the incomprehensible nature of Apple's decision to sync an iTunes account with a device instead of an account, which means you're liable to lose all of your music and data if you don't have it backed up.

Either way, if you're used to it then it's your best source. On the phone, Apple has an inbuilt app called Podcasts that is just the feature from the desktop version of iTunes built out on the phone. The app works quite well, and you can find good podcasts using the search function. However, you need to know what you're looking for since it is a library and not a recommendation machine.

Spotify

Spotify scores higher thanks to its recommendation engine, which will expose you to a variety of great podcasts along the lines of the ones you're already listening to. Most people use Spotify for music, so using it to listen to a podcast is not that much of a stretch. The only downside is that Spotify might not be available wherever you travel and is restricted in certain countries.

Soundcloud

Soundcloud is a podcast repository, which means that people upload their podcasts on here much like they do on Spotify and iTunes. The downside is again the recommendation engine, and you're better off knowing what you want to listen to prior to searching for it. The app is easy to use and works well on the desktop as well. Soundcloud is a no frills app, and if you like this sort of thing, it might be a good resource for you to use.

Google Podcasts

This is an app that exists solely to help you find the best podcasts. Given Google's ability in the search engine and data-tracking game, you're assured to find stuff that will keep you engaged. The best part is the easy integration with your existing Android phone. You can use Google Assistant's features to find what you need quickly and discover new shows.

You can also control playback and load the latest episode using voice control, so all in all, this is a great app to use. Best of all, it's free.

Castbox

Castbox is one of the best podcatching apps. A podcatcher is an app that hosts numerous podcasts. So iTunes is a podcatcher as is Google Podcasts. Either way, Castbox is the recipient of a number of awards from Google and offers the cleanest interface of all the apps. The best part is that it allows you to like, share, and comment on podcasts within the app itself.

This is a particularly attractive feature for podcast creators since engagement is usually done offsite. What I mean is that the app you upload your podcasts into will usually not have engagement stats on it. Think of it this way: When you upload a YouTube video, you get to see the likes, dislikes, and comments immediately below the video. This is not how it works for a podcast.

To generate this feedback, you need to post it on a web page or on YouTube and then use those platforms to generate feedback. The number of people who will travel off the podcast app to leave you a comment is low, so this reduces your feedback collection. Castbox makes all of this easy.

It is ad supported if you want it for free and the premium, no ad version costs $1 per month.

Pocket Casts

Unlike Castbox that requires a monthly fee if you choose premium features, Pocket Casts requires a one time $4 fee. This gives you full access to the app and all of its features. High on the list of desirable features is the notifications feature that lets you know when a new episode is ready.

In addition to this, it also allows you to stream your podcasts and not download it, which saves you space. From the podcast creator's view, the notifications and streaming ability makes this an attractive app to upload your podcasts to.

Stitcher

Stitcher is one of the biggest names in the podcatching space, but it also happens to be extremely buggy. The app has a free ad supported version and a no-ad paid version, but this doesn't improve overall experience. It has the basic features you need but beyond that, it's hard to see why one would choose it over the other apps.

I'm highlighting this here because of its huge library of shows. The app's recommendation engine by itself doesn't make too much sense, and you'll find some weird shows being recommended to you. Either way, Stitcher works as a basic app with no frills.

Overcast

Overcast is the best podcatcher for those using iOS. It works much like Castbox and Pocket Cast and curates shows for you. The premium version of the app is a bit expensive at $10, while the free version is supported by ads. The appeal of Overcast is restricted to iOS since it has been specially designed for this. The extensive library of shows will also ensure you'll always have something to listen to.

Chapter 2: Technical Aspects and Tools

I know you're anxious to get into the money-making stuff, but trust me, you need to understand some basic stuff about the technical aspects of podcasting before we get to all of that. Jump into the marketing side too soon and you'll find that your show is lacking the technical sophistication that bigger shows have and your results will suffer.

The technical aspects of your podcast begin with development and involve a number of things from initial prep work, finding a studio for editing, and uploading your audio. So let's begin by developing your podcast.

Development

Show development is the process of storyboarding. Hollywood types refer to this phase as the time when the gist of the show is developed and sold to studios. It is when these people figure out whether a project makes sense as a multi-season TV series or as a single movie or as a series of movies within a universe.

As you can imagine, there is a tendency for beginners to become overwhelmed at this point. There is indeed a lot of work that goes into researching the topic for your podcast. In fact, I've devoted all of the next chapter to dealing with the topic of niche research and topic selection, including tips on how to structure your show.

Technical Equipment

For now, let's assume you've selected a niche and the time has come to buy equipment for your studio. When I say studio, don't think of an elaborate setup with microphones and a completely soundproof booth. A studio can be something as simple as your office space or a place in your home where you will not be disturbed.

Audio technology has come a long way, and a lot of the equipment actually aids you in terms of cancelling out noise so you don't need to worry too much about how professional your setup is. So let's dive into the world of microphones and studio equipment. Remember, you can be as elaborate or as minimalist as you wish.

Some people get going with a single microphone, earphones, and a recording software. Alternatively, you can get technical with it and buy all kinds of rigs that will cancel noise and deliver a great experience. With this in mind, I'm giving you a complete guide to all the equipment that is available, and you can easily decide which ones make sense for you personally.

Microphones

The single most important piece of equipment you'll have is your microphone. No, you cannot use the mic in your awesome laptop. The fact is that the sound that is recorded by this mic is great when talking to your grandma, but you want your podcast to have a high degree of professionalism. Now, you don't need to have all sorts of fancy themes around it, but you do want to hit a basic level of production standards.

Before going around shopping for the best mics, you need to get familiar with some terms.

- Dynamic Mics - A dynamic mic is quite versatile in that it doesn't need an external power source. It has a wire coil and a magnet and generates all the electricity it needs. These are a great option for durability and they don't have too many moving parts, which makes them a breeze to use and set up.
- Condenser Mics - These mics require external power sources to work since they have a capacitor within them, and they rely on the electric current within the circuit to generate sound. They are quite sensitive, and the quality they produce is a lot higher than a dynamic mic. However, they are also less durable so there is a tradeoff.

When starting, you will be just fine with a dynamic microphone. While a condenser is a good investment, unless you already have a captive audience that you know you can monetize, you want to minimize your expenses as much as possible without sacrificing quality.

Pickup Patterns

A pickup pattern refers to the direction from which a mic will record noise coming into it. Generally speaking, pickup patterns fall into three categories: omnidirectional, unidirectional, and bidirectional. Let's look at these in more detail.

- Omnidirectional mics pickup sound from all directions as the name suggests. Generally speaking, these are unsuited for studio work since they'll have a lot of distractions within them and the resulting audio will be unfocused, no matter how good the quality is. Think of these as being the equivalent of the mic in your smartphone. It might produce great audio for the listener, but it picks up everything going on around the speaker. A large noise will drown out the speaker's voice. If you're outside a conventional, soundproofed studio, this will probably not work very well for you, or you will need to ensure absolute silence if using this at home.

- Unidirectional mics can be divided into two further classifications: Cardioid and Hypercardioid. As the name suggests, these mics record sound coming from a single direction. The problem with these mics is that while the sound is extremely focused, it can sound a bit unnatural and boomy if the speaker gets too close to the mic by mistake. A cardioid mic typically suffers from this quite a lot (Geoghegan & Klass, 2008). You need to be very cautious of the distance you maintain from it as a result. A hypercardioid mic or a shotgun mic has a very narrow range of sounds; it is tuned to pickup, and you need to be a very experienced voiceover artist to be able to pull it off. Generally, I do not recommend hypercardioid mics for home studio use or any sort of podcast use.
- Bidirectional mics don't find much use in podcasting since they pickup sounds on either side of the mic. If a podcast has two hosts who are dead set against using separate microphones, this works, but it's hard to find examples of this in real life.

So enough with the jargon. What you really want to know is which one you ought to buy. Well, I've made it easier for you by giving you recommendations in a series of budget ranges. This way you can match your budget to your preferred mic type. The best entry-level microphones are the Samson Q2U and the ATR2100.

Both of these retail under $100 and are very versatile. The most impressive fact about these choices is that the output can be transmitted as XLR or USB. XLR is what a professional studio uses to pipe sound into a mixer and create some magic with it. A USB is what you'll start off with, and you just plug it into your computer and off you go. This makes these mics very easy to use for talking on recorded Skype calls or interviews. Your audio is saved as a file and you can edit this in your software, which I'll talk about shortly.

One level up from these two we have the Rode Procaster, which is a dynamic microphone. To be honest, this is what I recommend you start off with since it does the job for podcasters who have large audiences as well. It retails currently for around $200-$230, which might sound like a lot, but consider that as your podcast grows you won't need to buy a new piece of equipment.

Another advantage is that you probably won't need to use a sound mixer because the output will be great by itself. Given that it is a dynamic mic, the recording will also be far more forgiving if you're recording in a home studio. The downside is that you will need a mount and a stand for this separately.

A lot of podcasters prefer to use the Blue Yeti, which is a condenser mic and plugs into USB directly. It will need an external power source being a condenser, but it comes with its own mount and stand so you don't need to invest in that. Best of all is its price, which is far cheaper than the Rode and retails at $100 or so. This makes it a very versatile mic for podcast use. Personally, I think it's a toss-up between a trade-off on sound quality versus the value for money aspect this microphone delivers.

A category above these two are professional microphones. These are the kinds of mics that professional recording studios use and are probably too much for you. However, if you're an audio geek and really want to get into it, the Heil PR40 is your best choice. It retails for around $400 and is a dynamic mic. However, as I said, it's probably overkill for your purposes.

When it comes to choosing your microphone, your budget is going to be the main point of concern, but remember that you should not sacrifice quality for price.

Pop Filters

Do you know what a plosive is? Well, I didn't either until I got into podcasting. Apparently, this refers to the consonants 'p' and 'b.' Don't know what a consonant is? Well, that's a topic for another book. The point I'm making is that when someone utters the letters p or b, a burst of air leaves their lips. This pops onto the microphone, and it results in a crackle.

To mitigate this, professionals use pop filters. These cushion the noises made and provide the listener with a more comfortable experience. There are two kinds of pop filters you can use: a foam filter or a screen.

A foam filter is the black hood that fits over a mic and is the filter that is most commonly used. A screen performs the same function but costs a little bit more and does a better job. For your purposes, a foam filter will do the job well.

Headphones

After your mic, your headphones are the most critical piece of equipment you will need to get right. Your headphones are your feedback in terms of what your audio sounds like, and you want to get them right. There are two broad options you have when it comes to headphones. You can choose either earbuds, which look inconspicuous, or the more elaborate noise-cancelling-type headphones.

There's no set pattern for this, but I personally prefer headphones over earbuds. They're just a lot more durable and provide better noise cancellation effects. Just like with microphones, you can purchase them at varying levels. Perhaps the most popular headphone for podcasters is the Audio-Technica ATH M30X, which retails for around $70. You could go for the higher end ones like Bose or Skullcandy as well.

The higher-end earbuds actually isolate your ear canal from sound better and are less cumbersome to carry around. However, you can expect to pay around $300 for what is still a very fragile piece of equipment. My suggestion is to save these for when you begin to make a large amount of income from your podcast.

One piece of equipment you should consider investing in is a headphone amplifier. Now this is not for your listeners' experience but for your own. The fact is that the amplifier in your laptop is not geared for optimal sound reproduction, and there is a possibility of you receiving incorrect feedback. If the sound on your show is of the absolute essence, then consider investing in this. To be frank, most podcasters will not need this.

Mixers

Mixers are equipment that most beginner podcasters will not need, to be honest. These are geared toward more professional shows. However, using a mixer can give your show a more sleek and professional feel. A mixer essentially allows you to mix sounds coming from various sources.

For example, if you want to introduce your show with a voiceover and then fade in background intro music, you can achieve this with a mixer. Otherwise, you'll have to pause recording, play your preferred music, start recording, and then edit the gaps out later before release. This is a pretty painful way of doing things, as you can imagine.

Another factor that might deter beginners away from mixers is that they can be complicated to connect to your computer. It's true that these days many mixers come with USB cords, but the fact is that most of them are designed for studio work and not podcasting. As a result, you need to have special cables and connectors, which I'll expand on in the next section.

If you do choose to go ahead and use a mixer, you need to watch out for the following things:

1. A good number of microphone inputs. For example, if you're going to be playing sounds from five sources, are there enough inputs?
2. Enough channel inserts - this is in case you use additional sound mixing equipment.
3. Quiet mic preamps - a fancy way of saying that the mixer raises the level of the audio signal from the mic.

I'm not going to spend too much time on the ins and outs of mixers since, frankly, I don't believe you need it as a beginner. However, if it's possible for you, do check out the differences between a professional sounding podcast with sleek intros and one from a smaller podcaster. Certain niches will demand additional investment as well thanks to the type of competition you'll have, so keep that in mind.

The best mixer you can get is the Behringer XENYX Q802USB. That is a mouthful to pronounce, but given that it sells for close to $90, it is a steal for the type of output it gives you. Furthermore, the fact that it comes with a USB makes it extremely practical to connect to your computer and use.

Bits and Bobs

This is the miscellaneous but important stuff you'll need to have to make your podcast a reality. Often, missing one of these little things will lead to delays and a huge level of annoyance. It's best to have these sorted prior to sitting down to record so as to minimize headaches.

The first type of cable you will need to know about is an XLR cable. If you're using a USB mic you won't need these. They're also referred to as microphone cables and have a male and female end. They connect your mic to other equipment and even connect two pieces of equipment to one another, such as a mixer to a recorder.

The next cable is referred to as the quarter inch, and this connects the mic to any other equipment like an amp or a mixer. Mixers will usually have what is called a combo port, which allows you to connect both a quarter inch as well as an XLR cable to it at once. Next up is the 3.5mm headphone jack that is found on headphones and earbuds.

There are different categories of these, and they can be differentiated on the basis of the markings and bands on their shaft. A plug or connector (Don't mistake a plug for an electrical plug, that's not what we're talking about here) with one band is called a tip sleeve or TS jack.

Two bands indicate a TRS or Tip Ring Sleeve, and a plug with two bands separated in the middle is called a Tip Ring Ring

Sleeve or TRRS jack. So what is the point of all of this jargon? Well, a TS jack feeds the audio to the source in one go. This is called mono transmission.

A TRS jack feeds it in what is called stereo. This means the left channel is being fed to the left side of the source and the right channel is being fed to the right. Think of it this way: If you're listening to a stereo delivery and remove one earbud from your ear, you'll only hear sound that is being fed to the remaining earbud. So, if music is being pumped on the left and dialogue on the right and you remove the left, all you'll hear is dialogue. In the case of a mono delivery, you'll hear both sounds equally in both earbuds. The TRRS adds a video component to delivery as well and is a recent addition.

Next we have RCA connectors, which are also called phone cables. Mixers usually have these ports on them, and you can connect any audio source using these. Lastly, you'll want to use a headphone splitter for situations where multiple guests are involved in a single conversation. If you don't use a splitter, the third person might not be able to hear what the second person said. A basic 3.5 mm cable will usually do the trick.

As a final word on cables, all I'll say is it is best to spend some money on them. There's no point connecting a $100 mic to a $100 mixer using a $5 cable. I'm all for minimizing your spending as much as is logical on studio equipment, but when it comes to these annoying little cables, it's best to splurge a little. I'm not saying you need to spend a fortune on them, but buy good quality ones depending on your needs.

Do note that if you're using a USB mic and headphones and are not using a mixer, you probably don't need these cables.

Audio Recording

Your audio recording software is the thing you'll be using to put together all of your recorded audio into a single, hopefully comprehensible show. There are two pieces of jargon you need to understand before proceeding: Double tracking and Multitracking.

Double tracking refers to the software's ability to record dual channels, one for the left and the other for the right, whereas multitracking is what enables you to blend music into the background and add it on top of your audio. These days, most software have the ability to perform both actions. When editing your sound, make sure you opt for the multitrack option if this isn't the default setting in the software you use.

GarageBand

For Mac users, it doesn't get much better than GarageBand. This is a free software that is specifically designed for the Mac, and you can edit and paste audio clips much like you would copy and paste pieces of images in Paint. While it won't suit a professional recording artist, for podcast clarity purposes, this will be more than enough.

You can separate and create multiple tracks for audio and other sounds and either blend them into one another or fade one in and the other out and so on. While there is a learning curve, it's not very difficult to master for a beginner.

Audacity

Audacity is an open source and completely free software you can use to remaster your tracks. Mastering is the process of editing and producing a finished audio track. Remastering simply refers to the process of processing the raw audio into a finished track or editing a previous version of it.

The software works much like GarageBand does, and there is a learning curve to it. It offers a lot of options for more advanced users, so beginners might be thrown off by this a little. However, you won't need to use most of it for your purposes. There are a number of tutorials online to help you figure out how to perform basic tasks. Once you make money on your podcast, you can hire a freelancer on Fiverr to perform more sophisticated editing tasks on your audio.

Call Recording

You will be conducting interviews with guests at some point, and you will need to record the conversation. The preferred VOIP software for these purposes is Skype, and the audio quality is pretty great. You won't need to remaster things too much unless your guest has a terrible microphone on their end.

The best software to do this is Ecamm recorder for Mac and Pamela for Windows. These software will record the audio and the video so you can use both feeds.

For now, this brings to a close our look at one aspect of the technical side of podcasting. The other side, in case you're wondering, deals with actually recording the audio and editing it for the best results. If I throw too much technical stuff at you upfront, you're likely to doze off, so let's take a break from all of this stuff and look at something else that is critical for your podcast: Niche selection.

Chapter 3: Planning Your Podcast

This is going to be a meaty chapter, so buckle up! Your podcast's niche or topic is absolutely crucial if you're going to make a splash and make some money. If you're an existing business and are looking to increase your sales avenues, then you might think you have a niche selected already. Well, not quite.

Think back to the first chapter where we saw companies like GE and Blue Apron use podcasts to popularize their products. The aim is not to publish something generic but to actually weave your products or services into the show in an entertaining way. If you're a freelancer who offers consulting services, creating a podcast about how awesome your services are is going to get you nowhere.

Niche selection involves understanding the intersection between what people want to listen to and how your product or service can be pitched to them as a solution. You could even use a podcast as a free method of increasing brand awareness. For example, if you find that people love listening to a fictional comedy show about an undertaker who is facing competition in his business in a tiny village, then why not produce the show and have it 'sponsored' by you?

I'm exaggerating, of course, but my point is that you need to remove the blinders and look beyond your industry. Podcasting is a varied space and you should cast your net wide. The first step is to determine how wide you want to go.

Niches

Here's the thing: The best-performing podcasts are true crime podcasts or investigative reporting ones. These are basically TV shows adapted for the audio format and are hosted by seasoned media professionals. Then there are the 'weird' shows, which are hosted by venture-capital-backed podcast houses such as Gimlet Media. You really can go off the deep end trying to get artsy with your podcast topics and structure.

If you have a product and already know your niche, then all you need to worry about is finding the related topics that enhance your product. For example, if you're an e-Commerce store that sells eco-friendly swimwear, then hosting an investigative reporting podcast about sea pollution makes sense. If your budget is tighter, then creating fiction stories for kids and 'sponsoring' the show will get entire families buying your product and will increase your brand awareness. The idea is you need to think of your niche as being something that is like a cloud of topics. Your product is one topic that is connected to a number of others.

Figure out what the cloud is and connect the threads to come up with a topic for your podcast. It could be fictional or nonfictional. In case you're wondering how on earth you're going to come up with stories to satisfy your audience, well, hire a ghostwriter from a place like The Urban Writers and have this narrated. Simple!

For those who have no clue how to start and need to find a niche, well, let's get into it!

Content Creation

As I mentioned earlier, your podcast is not a stand-alone business. You need to integrate it into a larger network of content creation that covers different formats. You need to have an active show page or bigger blog, at least one relevant social media account, and a YouTube channel where you can post video recordings of your podcasts and other informational videos.

So when it comes to niche selection, you want to figure out all the moving parts that go with this strategy. The ultimate aim of creating a podcast is to monetize it. Unless you're working for a nonprofit, there's no point in wasting your time creating something that is not going to translate into money for you. My point is that if you find a topic that is going to translate into a lot of listeners as a podcast but isn't going to lead to people buying your stuff from your website or blog, then there's no point in doing it.

The truth is that podcasts work brilliantly to increase brand awareness (Geoghegan & Klass, 2008). However, awareness doesn't always translate to sales unless you're big enough to have enough sales outlets. A company like McDonalds, which is present in every corner of town where terrible food is available, can afford to throw some money down to increase brand awareness. However, not everyone can.

Thus, you need to get your monetization strategy correct right off the bat. The best way to do this is to have multiple content channels and a niche where the audience is ready to spend money. At this point, most gurus will tell you to go away and buy fancy keyword research tools, but the truth is that with all of the updates Google has been releasing (the most recent one being B.E.R.T), these tools are becoming less relevant by the day.

The best place to start with niche research is to begin with yourself. What are your interests and what can you talk about for hours on end? Most of your thoughts will be toward broad-based niches, such as sports or movies and pop culture. Well, these would have done just fine when the internet was starting out. These days, you want to niche down farther to attract an audience. So while you may love to talk about the NFL, it's far better to niche down and focus on your local team or even a particular few players and positions.

Another great way of uncovering niches is to ask yourself what problem did you face recently and what did you do to fix it? For example, were you looking to buy a new beard balm and comb for yourself or for your partner? Well, what did you search? What gaps did you find? A gap is simply a piece of information you were looking for but could not find a clear-cut answer to.

A good example of a gap is the debt solutions market. There is always a gap here thanks to the ever-changing nature of laws and the political climate. While personal experience in such an area is not something you want to have, it is perfectly possible to perform research and deliver value to your audience. To top it all off, debt solutions affiliate programs offer fantastic payouts for both leads and purchases, so there's good potential for monetization.

Other examples include legal or financial niches, although personal finance is an extremely saturated niche and there's a lot of competition. We'll skirt over the competition for now and address it shortly. What you need to do at this moment is to brainstorm a bunch of ideas and come up with a list of at least ten niche topics you think you can become an authority in. Becoming an authority doesn't mean you need to be an expert.

It means that you can explore it, conduct research, and deliver good content. Beginners ignore the fact that a lot of authority sites on the internet started off in exploratory fashion and learned their field as they went along. So there's no need for you to have anything other than a high level of interest in that particular field.

Once you have a list, the next step is to dive in and look at keywords.

Keyword Research

Keyword and niche research often blend into one another. You can think of keywords research as being the process by which you flesh the niche out further and dive deeper into it. More often than not, beginners to internet content creation will go away and start writing about stuff they know about, thinking it's great content. Well, the internet is not the real world.

The internet and the way the big companies like Google, Facebook and Amazon are structured leads to people getting what they want, whether it's right for them or not. What I mean is that unless someone already knows what the best solution is to a problem, Google and the others will not give it to them simply because they can't.

For example, let's say someone is suffering from a condition that requires medication A when really medication B is what will help them. Unless this person willingly asks the question "Which is better A or B?" or "Is A the right choice for me?" Google will not provide content that answers these questions. If they simply search "Where can I find A and why does A work?" Google will give them the answers to this question.

If you write a blog post that will really help this person and will help them recognize that it is B and not A that will suit them, they stand no chance of finding you since Google is not designed to provide alternative viewpoints. Instead, it simply provides what the user wants to see. The philosophical conclusions of this aside, what this means is that what you think of as being good content might not match what your audience thinks of as being good content. In Google's eyes, it is the audience and not you that matters.

Hence, you need to focus on finding keywords your audience is searching. The first step to doing this is to login to the Google Keyword Tool that is available via Google Ads. This used to be free but in typical big company fashion, isn't anymore. However, it isn't too expensive. You need to spend a few cents on Google Ads and the tool is fully available for your use. It is worth the expense because other keyword research tools give you the same data and charge you a lot more (Booth, 2019).

Now, you must understand that Google has a mountain of data related to pretty much everything to do with human beings. You can go into the keyword research tool and type in your broad niche topics name. This gives you a list of keywords that are related to the niche. What you want to do is pick the long-tail keywords.

A long-tail keyword is a search string that contains more than two or three words within it. For example "best makeup for women" is a broad search term. I know it has more than three words in it, but the topic itself is broad. Consider the various categories within the term 'makeup.' You have mascara, nail polish, blush, eye liners, brushes, combs, hairsprays, etc.

The search string "best mascara for women in college" is a long-tail keyword. The focus of this search is far more specific. The person entering this is looking for mascara and they are in college. Also, they are looking for something for women, as the term highlights. The key distinction between a long-tail and a broad search term is the degree of specificity it contains. So don't simply go by the number of words in the search phrase.

Google will give you a list of keywords, both broad and long-tail. It will also give you the estimated searches per month. Here's the thing. Google obscures this data deliberately (Booth, 2019). So while a high number of searches is desirable, a better way of figuring out whether the term is searched for is to type it into Google directly. I'll get to this in a second.

Within the keyword tool, look at the results in the 'competition' column. If you see a bunch of 'high's in there, congratulations! High competition indicates that advertisers are large in number on that keyword and a glance at the estimated bid amount will give you an idea as to how much they're paying to advertise for that keyword. A higher amount means it's a pretty popular niche.

Think about it: Why would an advertiser pay top dollar for bidding on keywords unless they know they can make at least four or five times that as margin on a sale? This indicates a high profit potential for you. Note down the list of long-tail keywords that the tool gives you. Now, it's time to vet these.

Competition and Context

This is the toughest bit, but it is here where you'll create the framework of your content around the keyword. Start typing the keyword into Google's search box. Don't paste it in directly, but type it in one by one. See where it ranks in terms of Google's search suggestions. If you find that Google is suggesting it, then this is a term that is searched by enough people to warrant content. I suggest doing this in Incognito mode and after clearing your cache since Google personalizes content.

Note the other suggestions that Google gives you as well in the drop-down. In the results page, scroll to the bottom and note the other keywords in the suggested searches section since these are what people search for as well. Now comes the tough bit.

You need to evaluate the results on the first page. One way of doing this is to install an extension called Moz for Chrome, which will give you a bunch of metrics related to each result below its listing. There will be three or four of them, but the one you want to look at is the 'DA.' This stands for Domain Authority. Moz recommends that a website with a DA of less than 30 is a weak website, and you stand a good chance of ranking above this content.

Now, this is a hack and we don't really know what goes into the calculation of Moz's domain authority. The reason I say this is because Google doesn't provide its data to anyone. So Moz is estimating this as is every other search tool like LongtailPro, SEMrush, Ahrefs, and so on. At best, you can use the DA metric as a complementary one.

The real thing to analyze is the quality of the search results. If you search for "best mascara for women in college" and you receive a bunch of results that are articles whose theme is the best makeup options for women, the cheapest ones, the most environmentally friendly ones, and so on along with a bunch of Facebook pages and Quora questions, you've found a goldmine.

You see, social media websites do not have high authority in terms of Google's search results. It displays these results only if there's nothing credible out there. As for the other results, even if the DA is off the charts, none of the articles address the question specifically. There is a gap you can exploit. You can create a lengthy article discussing the best mascara (and only mascara) for college aged women and go deep into the topic. Odds are that over time, Google will promote your website to the first spot for this keyword.

I'll discuss the ins and outs of SEO later, but this is how you need to evaluate what content to produce and whether you can outrank your competition.

Across Platforms

You might be wondering what the relevance of all this Google-centric search is when you will be launching a podcast. Well, where do you think people are going to search for the topics you're going to talk about initially? Furthermore, how will you keep gaining new listeners without directing them to your website where you can inform them that you have a show in the first place?

This is why it's important to begin with search engine results. As added validation of your keywords, you'll want to hop onto YouTube and look at the search results you get there. Generally speaking, YouTube's algorithm works in the same manner as Google's does but there's far less competition on the video platform.

If you see just a few accounts with a large number of subscribers in the first few results, then this is a good topic for you to explore. If the entirety of the first page is full of large YouTuber accounts, then it might not be best suited for video. This doesn't mean you don't release anything in this keyword, but don't expect it to do brilliantly on organic YouTube search. Remember, Google takes precedence.

Next, you want to head over to iTunes and look at the existing podcasts in your niche. You can't do keyword research on a podcast since they don't have such tags, but you can look at the broad-based competition you will face. Remember, you can niche down farther than the bigger podcasts and still have a good following, so worry only if there are huge shows in your niche. You don't want to see ten shows with a following over a million listeners in your niche. For example, self-development and true crime see these kinds of numbers.

A good idea is to also hop onto Amazon and search for products related to your niche or any products you wish to market. Amazon's search bar will give you a number of results for the topics, and you should be able to find the products easily. How do you figure out their profitability, though? Well, Amazon gives you this for free via the Best Selling Rank or BSR.

The lower the BSR is, the more the product sells. You can go to Jungle Scout's website and search for the BSR to sales estimator and get an idea of the sort of unit sales a particular BSR in a category implies. It's not an exact number, but it is in the ballpark. This exercise will give you a good idea of the monetization potential of your niche as well as clarifying which strategy might be best for you to pursue. There are many ways you can monetize a niche, so doing this upfront is crucial. I'll cover monetization strategies later in this book, but it should always be near the top of your mind.

So you've found a bunch of topics people are searching for, you've noticed that there doesn't seem to be too much competition for these searches, and you're confident you can produce valuable content. Now, you need to create a website for you show, or create a portal where your podcast is just one element with which you can interact with your listeners.

Doing this is very easy with Wordpress, and my recommendation is to put up at least twenty posts on the topics you've researched before launching your podcast. It is ideal to have some content on your YouTube channel as well to spread the word about your brand.

All of this is in the future, though. For now, you need to plan your show.

Show Planning and Formats

The first step you need to take is to create a listener persona. This is the personification of your listener, and you need to determine what their tastes are and what kind of content they're likely to listen to. Where do they shop and work? Which other shows do they like listening to? Develop this ideal listener in your mind and a lot of your issues surrounding the format of your show will work themselves out.

A good idea is to take a look at your competitors. The internet is big enough that podcasts will exist for every niche even if these shows have just one listener tuning in. Listen to their content and try to see what their weaknesses are. In addition, doing this will also help you figure out the standards you need to make your podcast successful. Do you need fancy production gimmicks when your competition is a guy reading a bunch of text in a monotone voice, unironically? Probably not.

Next, give them a reason to tune in. What this means is you need to think of ideas for multiple episodes. This is why producing blog posts beforehand is such a good idea. You have topics to talk about already. Some of these topics can be merged into a single show, or you could break one huge post into multiple shows; it's up to you. Don't think of it in terms of extending your episode count. Always keep quality and value paramount. It's just good business.

Now you need to name your podcast. If you've chosen a snappy name for your website, then your problem is solved. Usually you can keep the same name as your website or blog. I recommend picking something memorable and to the point. Easier said than done, right? Well, I might as well let you know that you're going to spend a lot of time on this step. Like, a lot. You can use name generators that are free online and brainstorm ideas. Take your time with this since it is your brand.

Length

A considerable decision to make at this point is to decide how long your episodes ought to be. According to Stitcher, the ideal episode, as measured by user interaction, is between twenty to forty-five minutes long (Moazzez, 2018). You can aim to go over this, but when starting off I don't recommend it. I'll expand on this shortly.

You must also remember that your recorded content's length is not going to be your finished audio length thanks to editing. If you wish to produce forty minutes of finished audio, you're looking at raw audio of around fifty minutes to an hour. This sounds like a lot, but remember that the spoken word lasts for longer than the written one.

What I'm saying is that it takes you a few seconds to read a paragraph, but speaking it clearly and with the right pauses is going to take much longer than that, so don't worry about not having enough material.

Cover Art

The cover art of your show is extremely important. People will look at your show in their search results and if your cover art sucks, they're likely to click something else. Hire a good designer on Fiverr to do this for you. A good idea is to hire an eBook cover designer since self-publishers have the same kind of need as podcast creators.

They need their books to stand out in a sea of results as well, so you'll find that a good eBook designer will bring quite a lot to the table and might even design something that will look at lot different, in a good way, from what a podcast cover designer might produce.

Format

This is the big one. Which format should you choose? As much as you might love your voice, your listeners are going to get bored of hearing it over and over again and will require a change of pace. Shows with a single narrator that do well tend to exist in the topics of history and philosophy that aren't exactly geared toward monetization. You'll be operating in a topic where people are likely looking for advice, so you will have to operate in a format that allows this.

My advice is this: Combine all formats and get feedback from your listeners as to which one works best. You could do different episodes with different formats or incorporate elements of all formats into one show. Here are the different formats, other than the one-person solo show:

- Dual hosts – Two-host podcasts are quite common, but the downside is that you won't have complete control over creative decisions or earnings. So you will need to consider the pros and cons of this, not to mention how much you trust your business partner.
- Interviews - This is a popular format where you interview people who are successful in your niche. The upside is that your listeners will gain a lot of value. The downside is that it can be difficult to keep people coming back to you because they won't get a full sense of your own personality.
- Roundtables - This is a mega interview where you host multiple guests.
- Talk Show - This one is my favorite since it builds a sense of community and aids your branding. These shows are longer and incorporate all elements of other formats. You can start off with an opening monologue that can be a rant, have an interview in between, and then round it off with listener questions and advice. There's something in it for everyone.

Start off with a solo show but get into the talk show format as soon as possible by finding guests for your show. I'll show you how to find guests in the marketing chapters of this book. The talk show format also solves your problem of not having enough content to work with.

With three separate formats to incorporate, you'll have enough to talk about for at least a half hour.

Release Frequency

This is an important section, so pay attention. While you will be employing methods to gain traffic from Google and other sources, you need to focus on gaining attention from the algorithms of your podcast hosts. This means if you upload your podcast to iTunes, Stitcher and Overcast, you want to end up being featured in the "new and interesting sections" as much as possible.

There's two ways of guaranteeing this, and I recommend doing both. The first involves interviewing guests, which I'll explain in the marketing chapters. The other way is to release content as much as possible. This means you should aim to release your content at least every day.

If this sounds daunting, then that's good. You recognize the challenge ahead of you. This is why I recommended creating content ahead of time on your blog and YouTube channel so that you have things to draw from. A thing to note is that you don't need to release content of the same length or format all the time.

The algorithms don't pay as much attention to the length of your show as much as the activity. I'm not saying you release five minutes of blank audio, but you can do follow ups to your previous shows or host a smaller show of a different format. As an example, you can schedule one big show per week of forty minutes, and during the remaining days, you can release smaller shows twenty-minutes long.

These smaller shows can be solo shows or additional content from interviews you conducted or reaction shows that highlight the feedback you received from them. If you have any questions on your blog or YouTube channel about the topic, you can answer them here.

By releasing new content, you'll show up in the hot new shows category, and this will bring more traffic to you from the platforms themselves.

Scripting and Music

Scripting refers to you creating a script for your show. The reason you need to do this is because you want to avoid coming across as someone who's reading a bunch of text into a mic. In the beginning, this is going to be very tough for you to pull off. You see, a podcast ought to sound as if you're having a conversation with someone.

This is why dual-host podcasts work well; because it's easier for the hosts to carry on a conversation with each other. As a single host, you want to imagine yourself talking to your listener persona, but you'll also have to anticipate questions from them. A good exercise to do is to take a look at talk show hosts and watch how they conduct their opening monologues.

They're talking to an audience who they can see, but there's a few good lessons for you to gain from this. The first is that it is not a good idea to simply read your blog posts out loud. The written word does not translate well as a conversation, and unless you are quoting from books, you should not sound as if you're reciting an audiobook. Another good practice to follow is to develop a framework of a script and not each and every line of it. If you write every line in advance, you're again going to come across as someone who is reading something.

You want to sound as if you're talking to someone. This means you need to give yourself some room to talk and carry out an organic conversation as much as possible. Map out a few key themes and topics you wish to hit, observe other solo hosts to see how they do it, and figure out how you can make it work in your own style.

Music is another area you should spend some time on. Your show's opening and closing music needs to be in tune with your theme. Having blaring hip hop music for a show that's geared toward stay-at-home moms is mildly inappropriate to say the least. You can find great royalty-free music at Incompetech.

You can also shell out a monthly fee for over 100,000 tracks from the likes of Audioblocks. If you really want to go wild with it, you can use a custom service like the ones offered by Music Radio Creative. These companies will also create custom introduction and closing voiceovers.

Keywords

The keywords I'm talking about here are different from the ones we looked at earlier. These keywords are the ones you'll be submitting to your podcast host (iTunes, Stitcher etc). When people search these keywords, they'll potentially find you, although it depends on the competition in your niche.

So if you're creating an investing podcast, your keywords can include terms like investing, rich, money, make money, success, and so on. The trick is to have these keywords worked into the title of your show. Generally speaking, the greater the number of keywords you have in there, the better.

However, you do not want to stuff keywords into your title and have it seem nonsensical. So aim to hit a middle ground here. You don't need a hugely memorable name for your show unlike your brand. It could be something like "Get Rich With Value Investing - the Value Crusader Way." I haven't spent much time creating this title, so it could certainly be improved.

In this title "Value Crusader" is your brand name (if that doesn't appeal to you, how about "Value Saracen?"). This is what you need to name your blog and YouTube channel. Include keywords in your episode titles and get as many people to subscribe. This is straying into marketing so I'll address this later.

Once you've taken care of all of this, you just need to edit and master your audio.

Chapter 4: Uploading and Assembly

You might think that once you've finished recording, editing is just a question of a few points and clicks and you're done. Well, it's not that simple. On one hand, you could pay for a tool such as Alitu, which will automatically clean your audio and attach the relevant tags and so on. However, this costs money, and you want to keep your expenses low when starting out.

As a result, you will need to edit your audio manually, and you're going to run into a world of problems. As your podcasting guide, warning you of these is the least I can do. First off is the question of how much editing is good? You might want to preserve as much original atmosphere as possible to give your show an authentic feel. You might equally want it to be highly polished as well.

You'll have to play around with this and a lot of things, and this chapter is going to walk you through some of these issues.

Digital Editing

Thankfully for you, the process and mindset behind audio and video editing is largely the same, even in terms of how it works on software. I'm going to assume you'll use Audacity to record and edit your sound files, but really, pretty much every editor follows the same principles. The first thing you want to figure out is the feel of your show. This, of course, depends on your niche.

Your show format also dictates what your show should feel like. If you're in the undertaker niche and are talking about coffins, no matter how much money you can make, adopting a gleeful attitude is not going to be appropriate. Similarly, if you come across as someone who is low on energy in the beauty or health niche, you're not going to do very well at all.

Editing is really about monitoring the level of spontaneity in your show. Here's a simple mindset to adopt: You will make mistakes at first and you'll get a whole lot better at it as time goes on. So don't worry about making mistakes in the editing process and being less than slick with it.

Another attitude to guard against is perfectionism. Yes, it's your voice and yes it's your brand, but remember that your focus is to release content, not labor over editing it until it's perfect. The fact is that your listeners care about the content of your words, not how slickly you utter them. Unless everyone else in your niche is producing stuff at a very high level, you don't need to sweat this at all.

Think of podcasting as being the self-publishing version of radio. Your listeners do not expect to hear a perfectly produced show. They're looking for a person just like themselves who is trying to figure out solutions to issues they've had. If you're a company, you want to invest a bit more into this, but even here what I would suggest is to adopt a personal name for the show and tag your company's name as a 'sponsor.'

I mentioned the example of an e-Commerce swimwear company earlier. So your show could be named "The Clean Ocean Show with Cole Fugazy." During your introduction, you could mention your company's name as being the sponsor or as being the company associated with your podcast.

Simple. This gives you the best of both worlds. You won't have listeners expecting a corporate show, and you can still direct traffic to your products.

A good idea to implement is to track the amount of time you're spending editing. Usually editing takes a while, but if you find that you're taking four hours to edit and compress your raw audio into a thirty-minute show, you're doing something wrong. Aim for a 2X multiple on your time. So every minute of finished audio requires two minutes of editing. A half hour show should not need more than an hour's worth of editing.

Remember above all else that your podcast is a conversation. A few stray 'ums' and 'ahs' are perfectly acceptable.

Types of Editing Software

All sound editing software falls into roughly two categories with a few hybrids here and there. These categories have to do with how the sound files are edited. Software such as Audacity allow you to make changes directly to the file itself. In other words, when you edit the file, your original recording is lost the minute you save the new changes.

This is not always a good idea when you're starting out since you will make mistakes. As a rule of thumb, do not ever delete or overwrite the original file since you never know when you might need it. A better method of handling the situation is to make a copy of the original file and edit that instead. This way you won't be tampering with your raw audio.

Some software will do this for you automatically. In fact, all of the higher-end editing software does this as standard. You'll import what is called a reference to your audio file into the software and will then edit that. The software saves the edited reference as a new file.

This might not seem like a huge thing right now, but trust me, when you screw up your first few edits, you'll wish you had your original audio back. So make sure you understand how your software works prior to fooling around with it and also edit copies of the file if the software doesn't import a reference to the original.

Editing

One you've uploaded the file, you'll need to proceed with the editing. Generally speaking, it's a good idea to start by removing any offending silences in the recording. This is because these are the easiest to remove. The obvious silence to remove is at the beginning of your recording.

Don't trim too much of the silences in between since this will lead to your show feeling a bit too sterile. When people talk, they pause and take a breath and so on. So you want your audio to come across as being as natural as possible. As you can imagine, there's no template I can give you to edit your audio; it's entirely up to you.

Editing is pretty simple in all these software programs. You simply highlight the offending portion of the sound (represented by a waveform) and then delete, copy or cut this portion as you wish. You can transpose these segments elsewhere or remove them entirely. Again, keep it simple and focus on getting your audio to air as soon and as easily as possible. Your listeners are not expecting a super slick show, so don't worry about this.

Poor Audio Quality

The worst thing that can happen to a podcaster is that they conduct an interview with a guest and then find that for whatever reason, the sound quality is terrible. Calling them up and asking for another interview should be your last and nuclear option. Do this only if it is unavoidable. For now, what you can do is play around with what is called the notch filter in your software. The location of this filter is in different places depending on the software, so I can't tell you exactly where it is. However, a simple search for your software's features will highlight this easily.

The notch filter is typically used to remove the hissing noise from older recordings. However, it also removes a lot of other audio disturbances such as crackling and other problems. It takes a while to play around with the settings, but you'll find that more often than not, it manages to clean your audio up very well. The only thing it cannot do is compensate for low volume, although the waveform editor should do the trick.

To learn about your software's notch filter and how to use it, you can go onto YouTube and search for a guide on how to use it. There is a lot of helpful content out there. If time is of the essence, simply hire a freelancer to edit your files and pay them extra to see how well they can clean it up.

Mixdowns and Mastering

A mixdown refers to the process of you adding music and other sound effects to your audio. Always edit prior to adding music. This is because once music is added, you'll find that any edits you make might disrupt the music and will sound odd to your listeners. Create the music tracks as a separate waveform and preview how your finished file will sound before compressing everything into a single file.

Once you've compressed it, there are a few more steps you can take. This process is called Mastering and it goes back to the days of vinyl records. These days, Mastering all about playing with the EQ settings and enhancing the overall quality of the sound. Truth be told, as long as you have a good mic and have chosen good quality audio and have made good edits, you won't need to worry too much about the more technical aspects of sound editing.

As a top tip, you can use music to make a lot of the ambient noise that occurs when you record at home. This is called using a music bed, which is a stream of low background music that keeps running. Obviously, you don't want it drowning out your voice, but consider using it if you find the background noise annoying or distracting.

Another tip I will pass on is to learn the keyboard shortcuts once you've edited a couple soundtracks for your show. Once you ditch the mouse, you'll find that the editing process can take as little as fifteen minutes (without including the process of listening to your final audio). You'll get better at it as you go on, so don't worry too much about this. Worst comes to worst, you can hand your files over to a freelancer who will do a good job of producing your final audio.

Tips

I'm just full of tips and tricks right now! A helpful thing you can do when recording your audio is to create what is called a punch list. I'm not sure of the etymology of the name, but it's what is used by podcasters. This is a list of edits that you note down to remove later during the process.

Start your editing process by creating a punch list. This will help you organize your editing sessions better, and you'll spend less time clicking around and wondering whether you ought to remove something. Another thing to remember is that the point of your editing is to direct the listener's focus as opposed to producing high-quality audio.

There's a difference in the two approaches. The former focuses on value, whereas the latter is far more concerned with vanity. Musicians and artists are concerned with the latter and that's fair enough, I guess. For your purposes, though, think of your audio as being a path along which you're guiding your listener. Adopting this approach will clear up a lot of questions you'll have where you'll wonder whether or not to leave something in or remove it.

A good technique to master is the usage of fades. Often what happens is that despite your best intentions, you'll be left with holes in your audio that you won't be able to fix or clean up. Pushing the next segment would result in an unnatural flow. What podcasters usually do in such times is fill the gap with soft ambient music. Alternatively, you could record something else, but this would require you to speak at the same timbre as the preceding bit of audio, which is tough to pull off.

This is why a crossfaded bit of ambient music helps cover the hole and it indicates to your listener that one topic is finished and another is starting. Keep these as simple as possible and don't run away with yourself. The same applies to adding intro and outro music and sound effects in between.

Uploading and Distribution

Here's a step that flummoxes beginners. Your audio is edited and you've gazed upon it with pride. After all, this is your creation! Now you need to get people to listen to it and start gaining value from it. The question is: Where should you upload it? There are many places you can listen to podcasts, so do you have to upload it to every single one of them?

That would obviously take forever. You can't spend an entire day simply uploading a podcast. After all, you do have a life to live. So let's take a look at how uploading works and the options you have.

Hosts and Directories

The host or podcast host is the place where your podcast will live. Much like how your website needs to be hosted by someone to store the data in it, your podcast needs a similar service. Your choice of podcast host is crucial if you want to achieve success with your show.

This is because the podcast host carries out a lot more functionality than a website host does. The podcast host creates what is called a feed, which is what you'll be submitting to the places where you find and listen to podcasts. In addition to this, the podcast host also configures your audio to best suit the formats of various directories (these happen to be the same for the most part).

Some podcast hosts also give you the option to create and host a website for your podcast within their platform itself. I don't recommend doing this. This isn't because the service is bad, but more due to the fact that you should be thinking of your podcast as a large cog in your content machine. For these purposes, you need a fully functional and robust website. Posting your podcast to your website is pretty simple, and I'll show you how shortly.

So which are the best podcast hosts? Well it comes down to what you're looking for. There are three choices which are universally rated as the best (Maclean, 2017). You can think of them as being in the bare bones category, the growth category, and the slightly expensive but good value for money category.

In the bare bones category, we have Buzzsprout. Buzzsprout is the easiest to get setup with, and you'll be up and running in no time. The best feature they have is the ability to generate a video preview. You simply select a section of your show to highlight and they'll automatically create a preview video (with a standard background) and embed your audio in it. This can be used to market your podcast.

They've also added the ability to add chapter markers. Ever seen those long YouTube videos that are timestamped? Well, this is the audio equivalent. That way, your audience can jump to a particular point in the podcast to listen to the topic they want. The best part is that Buzzsprout offers a free plan, but to be frank, I would not go with this.

The free plan supports two hours of content every month. Remember how I said you need to release content regularly in the first month to be featured on iTunes? Well, you'll run against this limit pretty soon. In addition to this, they'll also delete any files that are older than 90 days. You'll have the original audio, of course, but this means the directories you submitted your podcast to will no longer have the older episodes.

The good news is that the most expensive Buzzsprout gets is $24 per month. However, there's bad news here as well. You can upload up to 12 hours of new content per month. If your show is half an hour long, you can just about get away with it, if you skip weekend releases.

Buzzsprout offers pretty good analytics data and gives you a count of the number of listeners you currently have. This might not sound like much, but podcast analytics are hard to come by because there's no way of tracking anything in an audio file. Remember that this is a bare bones host. So you'll be limited to one podcast per account and you cannot have team members login and manage your podcast for you, which is a bummer once you grow.

For such situations, I recommend using Captivate. Now, you might be thinking that this is going to be expensive. Well, the priciest plan on Captivate will cost you $99 per month. However, this is based on the number of downloads you get. If your show exceeds 100,000 downloads per month, this is the price you pay.

Captivate has all the features of Buzzsprout and more. You'll pay $19 per month for up to 10,000 downloads, and to be frank, this is probably where you'll be for a long time. You should not have any problems with monetization above this many downloads, so I wouldn't worry about it. The best part is that you can host multiple podcasts with them for the same price. It entirely depends on the number of downloads you get.

It also allows you to delegate roles to team members and have them manage your podcast for you and gives you the ability to share your podcast on social media and publicize it.

A third option you have is to use Transistor.fm. This is a platform that is run by a couple of podcasters themselves, and a lot of the little niggles that you'll find on the other platforms are nonexistent here. The best part is that Transistor allows you to export your audio file to YouTube with the click of a button. So you can have a standard background and create a readymade YouTube video to promote your podcast there as well. A lot of podcast hosts make use of this option.

The pricing is the same as that of Captivate and works the same way. You can host multiple shows under one account and plan and team members can login and manage your show for you without any hassle.

Now that we've gone over the hosts you should sign up with, here are two to avoid. These are Soundcloud and Anchor. The platforms themselves aren't bad. In fact, they're free and this is precisely the issue. First off, Soundcloud is not designed for podcasts although you can host them on there. It is a platform to share music. As a result, the analytics the platform provides are close to nonexistent.

Analytics is also the issue with Anchor. It offers a seamless transfer to iTunes, but the issue is that your podcasts will be submitted through their own Apple account. iTunes provides a ton of useful analytics and you'll have zero access to any of this. So the bottom line is: avoid free podcast hosts, they're just not worth it.

There are many other great hosts out there such as Libsyn, which is the oldest host, Bluebrry.com, Soundcast and so on. You won't go wrong choosing any of these to be honest. You might be wondering whether you could host your podcast by yourself? Well, it is technically possible, but you'll definitely run out of bandwidth and your hosting costs will increase despite traffic being low. So it's not a great tradeoff.

What you can do is embed your podcast on your website. The way to do this is to choose the embed option in any host and this will give you a piece of HTML code which you can then paste into your website's backend and the podcast player will appear on the page. Common practice is to create a new blog post and embed it there. Below the player, make sure you paste the transcribed audio so that you can use Google to rank for keywords you use during your show.

So that covers podcast hosts. The question is which directories should you submit your shows to? I've covered a few of them previously such as iTunes (which is officially called Apple Podcasts by the way) and Stitcher and Overcast. In addition to the ones previously highlighted, I recommend submitting your feed to TuneIn Radio, which integrates with Alexa brilliantly.

Remember you submit your feed to these directories and not the audio itself. You upload the audio to your host's website and then click the button to create a feed and submit it. Some hosts will offer the choice to do it within the platform itself. To submit your feed to places like TuneIn, you may have to create a feed and manually submit it.

The directory will then constantly check for new content so you don't have to do anything after this initial step.

Chapter 5: Video Podcasts

Video podcasts or videocasts are actually a lot simpler than they sound to create. All you'll need in addition to your regular equipment is a screen recording software and a camera, which captures your face. In fact, you don't need this either because you can use software that records calls in HD.

This chapter is going to show you how easy it is to create a video podcast and why you need to be doing this if you're serious about growing your show.

Videocasting

Getting traction on the internet is all about getting as many eyeballs on your content as possible. To achieve this, you need to have as many channels as possible to disseminate your content from. This is why many podcast hosts offer the option to convert your audio into a video and upload it onto YouTube. However, the downside here is that you'll be posting a simple background with your voice over it.

This is fine if you're starting out, but if you really want to maximize your opportunity to grow, you need to have your face on it. This gives your show personality. The internet is an impersonal place and people are still not conditioned to buy stuff from someone who they do not know. This is why in-person selling still works. To get someone to part with their money for your product, you need to build an emotional connection.

Video is a great way to do this since it allows you to present yourself as a real person and not as a voice floating around on the internet. To achieve this, you need to follow some simple steps.

Equipment

Believe it or not, you don't need anything other than an HD webcam and a Skype call recording software. If your show is not going to comprise any interviews (which I do not recommend), all you'll need is an HD webcam. If your laptop is high end, you could use the inbuilt webcam, but these are usually a bit dodgy. An alternative is to record your podcast using your smartphone mounted on a tripod. Phones these days have pretty amazing cameras, so you can record this as a video on your phone and then save it to your computer.

You will need to figure out how your mic and headphone arrangement is going to work in such a scenario. You might need to use a wireless headset. However, it is a good investment and you'll save on carrying around an addition piece of external equipment like a webcam.

When conducting interviews, you can record the Skype call using software I've already highlighted previously, so this isn't a hurdle in any way. What you will need is a decent movie editor. Windows and Mac computers have good inbuilt movie editing software, and these will do for the most part. If you want to get nerdy with it you can choose software from Adobe.

Audacity and any voice editing software has the ability to rip just the audio from your video file so your sound editing process is not going to be affected. Once you've finished editing your video file using the software, all you have to do is upload it to YouTube and that's it! I'll talk about optimizing your videos for YouTube in a later chapter.

The bottom line is that it's pretty simple and straightforward to create a videocast and you don't need anything additional. So you should be doing this as a matter of default.

Fancy Videos

Given that videos involve participation of people, there is no end to the extent that you can make them look pretty. There are a lot of elements that go into producing a great video, and things like lighting, choice of clothes, scripts, sets, and so on can be fiddled with.

It is up to you to decide how you want to do this and whether it is important to your audience. There's nothing wrong in appearing as fancy as possible, of course, but you should keep your audience's needs in mind. Is their priority a fancy set or quality content? Take a look at what you competitors are doing and understand that you'll have to match them at the very least.

A good example of this is the SEO and SaaS marketing fields. The people who operate in this niche produce extremely slick videos that you wouldn't expect given the technically heavy topics. Also keep in mind that you're not only competing against podcasters in the niche but also against regular YouTubers. So research your niche thoroughly before jumping into it via video.

Once you grow in size you can hire a professional studio and a team to manage things like production and post-production. In essence, you'll be creating a TV show every week, and there's no way you can do this all by yourself if you want to remain sane. So start off small and don't overdo it. Video has great potential, but there's also the possibility of going off the deep end and spending too much time planning and not doing anything.

So explore your niche well for video potential and start optimizing your strategy for video from the first day. You'll save yourself a lot of headaches this way.

Chapter 6: Free Traffic - Part One

So, finally, we've come to the stuff you've been wanting to read about! From this point on, I'll be dealing solely with marketing and making money off your podcast. The technical stuff is all behind us now. The only thing I will ask you to recall and will make constant references to is the process of niche research.

So do take the time to refer to that since the activities you carried out during that phase are what you'll be repeating in a few cases. Besides, the concept of keywords and leveraging free traffic is what we'll be discussing and all of this stems from similar concepts as niche research.

Remember that your podcast is just one piece of your content machine. It might be the biggest piece, and it could be the only piece you monetize. Either way, you need to leverage all of your potential content distribution channels to spread the word about your show.

So let's begin with the most basic form of content distribution: Search Engine Optimization.

SEO and Your Website

SEO has been around since the advent of the internet and is the most crucial piece of inbound marketing. The sad thing is that most beginners give up on SEO far too soon and don't put in the sustained effort that reaps them rewards in the long run. A common question that is often asked is how long will SEO take to work? The infuriating answer is "It depends."

There's no easy way to quantify how long it takes for this to work. It depends on your niche selection. However, I can give you some guidelines on how long it should take you to see results. If you don't see anything within this time frame, then you've gone wrong somewhere. Setting yourself up for SEO success begins right from the moment you choose your niche.

Remember how you looked up competition for your keywords? Well, the lower the quality of your competition, the sooner you'll be able to rank. Let's look at Google's basic philosophy, which will help you understand how to start getting organic traffic quickly.

Authority

To truly understand SEO, we need to travel all the way back to Google's mission statement. It was to "organize the world's information'" (Stokes, 2019). I say 'was' because these days the mission seems to be something to do with world domination, and the company has seemingly long left behind it's do-gooder past. Either way, the intent of the search engine is to organize information. Google views itself as a vast library and it aims to be able to direct people to the right shelves for information.

The books on these shelves need to be vetted. Think of it this way: If you're looking for stuff to do with the private life of JFK (the president not the airport), would you like your librarian to direct you to the unverified but salacious stories, or a verified and vetted biography? Well, this depends on your intent. More on this later.

For now, Google largely assumes that you're looking for stuff that has a high authority rating. So the gossip magazines from the '60s are out and verified biographies and stories from trusted sources are in. A few things to note at this point: First, the greater the number of authoritative sources, the more subjective Google's decision becomes.

Sticking to the same subject, if there are multiple credible news sources such as The New York Times, The Post, and a host of other credible newspapers and magazines such as TIME, Forbes, and so on, who's to say one is more credible than the other? Well, Google does, and it does so because there is no other option. Given the large number of credible sources, you'd imagine there would be many shelves containing all of these. So which shelf gets recommended by the librarian? Well, it's a crapshoot.

This is why you need to be sensitive to competition. The lesser the credibility of existing sources, the better it is for you. Websites/articles that don't address the problem directly and social media posts (which are classified as uninformed opinion by Google) are pretty low on authority. Thus, if you've done things as recommended, you should reduce the time it will take you to appear at the top of search results drastically.

The next thing to note about authority is to figure out how Google determines this. One obvious way is brand name. The NYT and Post are obviously high-authority sources no matter what the President's Twitter account says. Why are they considered high authority, though? Well, this is because a lot of people say so. You can make many intellectual arguments about this, but all of them boil down to this in the eyes of Google. Authority, according to them, boils down to whoever is making the most noise.

Noise equates to traffic and a low bounce rate on websites. A bounce is when a visitor clicks on a webpage and then navigates back to Google. They would do this if the results they received are not satisfactory. If you like what you see, you're going to stay. Another metric of 'noise' is to look at how many people are talking about a source.

The NYT is going to be quoted in many different websites and in other equally high-authority sources. This is called a backlink on the internet. A backlink is when a website links to another. Google looks at the quality of backlinks as a metric of authority. So if you're talking about the benefits of grooming your dogs and a larger website dedicated to pets links to you, this is pretty significant in Google's eyes.

If a website dealing with alcoholism links to you, it's not very relevant. If a website is known to spam its users and gift them free virus links to you, Google will actively penalize you (Stokes, 2019). Links placed in forums and comment sections of other blogs are automatically categorized as 'nofollow' links. This means Google will not consider this a valid backlink. So spamming blogs with links to your website will not work.

A large number of shares on social media is also another metric of high authority. Since 2015, Google has been implementing what is called an E-A-T guideline. EAT stands for Expertise, Authority, and Trustworthiness. All of these are various elements that go into the overall authority score of your website. The better you can prove this to Google, the more they'll promote you in the rankings.

Lessons

So with all of this in mind, what can we conclude for our aim of getting ranked as quickly as possible? Well, understand that quick in this context means around three to six months. You're not going to receive much organic traffic from Google within a week. That's just unrealistic to aim for. The way it works is this.

Google will crawl your website constantly and will take note of the keywords, topics, and niche. It then lets it marinate for a while as it correlates what you're writing about with its own vast library of keyword and search data. Once it develops a decent idea of what your website is all about, it begins to send you a little traffic now and then and monitors the bounce rate.

If the bounce rate is low, it progressively sends you more and more traffic until it reaches a point where the bounce rate is too high. It then recalibrates and adjusts the traffic it sends you. This is why SEO takes time and isn't an overnight process. However, the traffic if free and Google actively helps you ascend in the rankings because it cares about your content and how relevant their recommendation is to the user.

Since bounce rate is so critical, we need to structure our website content in such a manner that it minimizes this. Thus, while some of your posts will be your transcribed podcast, you need to surround it with other forms of informative content. So what kind of content is informative and useful?

Well, answers to questions, tutorials and tips and tricks are the three categories you need to focus on. Answers to questions-type posts are where you address common problem areas and give your reader a solution. You can create a show surrounding this topic as well and answer it on there. Your transcript can then fulfill this blog post requirement.

Tutorials are the really meaty and in-depth articles, kind of like how I'm diving deep into SEO here. I'm hitting a number of related topics to the subject, and if I were to run this into an article, I could go on for over 2,000 words or even 3,000. This is a good thing in Google's eyes since the length itself implies you're imparting a lot of value by going in-depth.

Tips and tricks is a catch-all segment. These are where you'll find listicles such as "Top 9 tips to groom your dog" or "The 7 things your dog thinks of when you give him a bath" and so on. These can be lighthearted and funny but also provide quick shots of value in a list format. At this point, I would like you to visit any blog you like (not a media house like HuffPost but a regular blogger) and you'll find that despite the article titles not seemingly matching these categories at all times, the content itself conforms to this framework.

Moving past content length, we arrive at your UX. UX stands for user experience and can be boiled down to the following question: How easy is it for users to find information on your website? Let's say someone likes what you've written and wants to ask you a question. How easy is it for them to leave a comment? If they want more personalized attention, how easy is it for them to send you an email?

How Google monitors this exactly is not known (Booth, 2019). However, UX plays a significant role in your search rankings. Your UI is what determines the quality of your UX. UI stands for user interface and is the design of your website. The long and short of all this is that you should not be a cheapskate and opt for a free theme just to save costs. Put your UX at the forefront. Keep in mind that UX also involves the quality of your content.

If your content is rubbish and people are bouncing, that's a sign of a bad user experience. This is why I mentioned back in niche selection: Stick to things you know you can provide quality content in. You can't make things up these days since Google is a lot smarter.

EAT means that you should look to establish your own and your website's authority as much as possible. What does this mean in practice? Well, to understand this better you need to first understand YMYL. This stands for "Your Money or Your Life," and no it isn't something I made up. Google themselves used this acronym in their search guidelines document released a few years ago (Booth, 2019).

YMYL refers to those topics which materially affect the user's health, wealth and any aspect of their well-being. For example, nutrition and fitness are good examples of topics that are YMYL. Personal finance is another topic which is YMYL. Google enforces EAT very strictly in these cases so if you're starting out in these niches, you're better off getting yourself some credentials.

This is why you'll notice a lot of websites these days make the author's bio and credentials very explicit when it comes to such topics. Popular websites such as Healthline make their author's credentials extremely explicit as opposed to their competitor WebMD, and as a result, Healthline ranks higher on the first page for pretty much every topic, even if WebMD makes the first few search results. This is EAT in action.

To be more precise, that's the E and the A in action. T (Trustworthiness) can be established in other ways. Your website's security certificate, usage of the https protocol, an accessible privacy policy, a physical location for your business/website, refunds and return policies and so on help Google figure out whether customers can trust you.

Another method of building trust is to cite relevant sources in your work to credible authors if you're sharing information, and having an author bio below your article to certify who you are. This applies to all topics, not just the YMYL ones. In non-YMYL topics, EAT is not as fundamentally relevant but the ethos of it remains.

Post Structure and Intent

To succeed at SEO, we've seen how you need to provide high quality content to your visitors. So how do you structure this as a repeatable framework? For starters, you need to have a headline that is relevant to the topic. This is a common sense thing, but you'd be surprised at how often people write one headline and then talk about something else.

So if you're releasing a show transcript where you spoke about tuning the engine of an old Jag, then you need to include the relevant keywords in the title. Also mention it somewhere in the first paragraph. Don't stuff your keywords. The bigger question is: How do you structure your article? If it is a podcast transcription, then breaking out your timestamps into subheadings is a good way to do this. This gives the reader a quick guide as to what it's all about.

However, the question still remains: Let's say you've found a keyword and now need to create an episode around it. How do you do this? Well, thankfully Google gives us all the tools we need. First, look at the searcher intent (which I briefly mentioned when talking about JFK previously).

Searcher intent is now more important than ever thanks to Google releasing its BERT update. I'm not going to bore you with what BERT stands for, but it basically means Google can tell what the intent behind a search is and its AI is improving at this at a fast clip. As an example of how this works, let's say you find a keyword called "natural soaps suitable for men," which you feel is underserved.

Prior to BERT, if you fired off a 1,000 word article that was just a listicle of the ten best natural soaps, Google would have ranked you pretty high and sent you traffic. Post-BERT, though, your article will be demoted in the rankings. Why? Well, what is the intent behind the search query?

Perhaps the searcher is looking for an answer as to whether natural soaps are good for men? This means they want to know what goes into them and why they're beneficial. They also want to know the different types of soaps and which ones have a masculine scent (guys don't want to smell like fruits or flowers). And, yes, they could do with a list of suitable soaps as well.

So really, what you want to be producing is a guide, not just a product recommendation list. Searcher intent is the key for your content moving forward. This is how you structure your show as well, and it will translate well into the written form too. A good method of looking at possible structures if you can't figure it out by yourself is to look at the top results for the query.

Combining the topics of two or three of them (assuming they're not all encompassing) is a great way to create valuable content. Another key thing to integrate into the structure of your shows and written posts is Latent Semantic Indexing or LSI. You're probably hungover now from all the acronyms I've been throwing your way, but bear with me.

LSI keywords are the related keywords to the one you're searching for. For example, an LSI keyword for "natural soaps suitable for men" would be "are natural soaps safe for men" and "who are natural soaps for." Remember the keyword cloud I mentioned back in the niche research section? Well, LSI keywords are a part of that cloud but are more like topics instead of specific keywords.

So how do you find this if you're new to a niche and are exploring it? Again, Google shows you the way! Have you noticed those questions and answers (People also ask - PAA) it displays within the search results? Well these are the LSI topics (Booth, 2019). This is what Google sees other people searching in addition to the keyword you searched for.

Incorporate a discussion of these into your show and post your transcript. Google will take notice, and you'll find traffic coming your way over time because you're hitting the topic from all sides. Not to mention the fact that LSI data tells you what your viewers find the most relevant and useful for discussion with regard to the topic. They'll appreciate you a lot more and you'll find your organic reach growing.

Another thing to note about PAA is that sometimes the list expands and sometimes it doesn't. The expanding PAA gives you a roadmap as to which angle to take. Once you click a particular question, the questions below it branch into that subcategory. So use this well to create smaller follow-up shows and topics to discuss later.

Backlinks

As I mentioned earlier, building backlinks is essential to establish some sort of authority. Now, there is a school of thought that believes that you should build backlinks proactively, but I'm not one of those people. If you ask me, if your niche is small and if there aren't too many competitors, you don't need to seek backlinks, they'll come to you eventually.

However, if you are in a crowded niche, you need to start building a backlink strategy right from your first post. If anything, it will serve you better to build backlinks prior to releasing content too much. The way to do this is to first create a few large posts which offer quality content.

What should the topics of these posts be? Well, you can create a long post that dives deep into a topic of interest in your area. Just to clarify, I'm using posts and podcast shows interchangeably here. Your post is simply your podcast's transcript. You can release your shows in a fast and furious manner, but hold off on releasing too many blog posts at the start if your niche is a crowded one.

What you want to do in these initial shows is to not just dive deep, but also use methods, recommendations, or techniques that have been proposed by other influencers in your niche. Use their names in your show and try out something they've recommended and highlight these results. Can you see the point of this?

Well, if you can't here it is: Everyone loves being mentioned somewhere, whether in a blog or in a podcast. It gives our egos a little boost, and we tend to look favorably upon someone who mentions us in a positive light. So if this person who mentioned you turns around and asks you to appear as a guest on their show, would you be willing to appear? Sure you would! Would you be willing to publicize this to your audience and post a link to their blog? Sure you would! Hello, backlinks!

This is how backlinks are built, and this is pretty much how you land guests to interview for your podcasts. There are other ways, of course, and I'll cover landing guests shortly. For backlinking, though, this is the best method. If you happen to be in a technical niche, you can author a study and then message the influencer requesting their feedback on it and give them a co-author credit. This way you get your name out there and the influencer gets to highlight an important, credibility-building study. They will willingly link back to your blog and mention your show.

Guest posting is the traditional method people seek to build backlinks, and it works. However, you need to be sensitive to the sort of content the other person expects. A good way to figure this out is to look at their top performing posts. Most blogs have a "most popular" section. Scan this to figure out which content works best on their blog. Take a look at their content and come up with something better or a twist on what they've written. Even better, invite them for a chat on your show.

This approach is likely to get you some pretty good backlinks back to your blog. Once you have even one link, start releasing more content along with the transcripts of your shows. Your audience will grow astronomically. The only downside is that there's a risk of people reading your blog and not downloading your podcast. To mitigate this, you can hide some content from the transcript. Don't be harsh with this, but highlight that the podcast has information that is not highlighted in the transcript. Alternatively, traffic is traffic and between your blog's reader-only traffic and your podcast's listeners, you can make a pretty good income. So you need not worry about which form of your content is being consumed as long as someone is consuming it somewhere.

Guest Spots

One of the best ways you can grow your audience is by convincing people to come on to your show as a guest. Keep in mind that the influence of your guests must be in line with where your show is at. If you've just started out an astronomy podcast and have a grand total of two downloads, Neil Degrasse Tyson is not going to be interested in you even if you were his favorite niece.

The key to finding good guests to book is to look at influencers who are a tier above you or equal to you. When you're starting out, there's not much point inviting people who are on the same level as you because, well, neither of you has an audience. However, once you've reached a decent size, collaborating on a show together will help you grow each other's audience sizes.

So let's take a look at how you can build your own credibility by interviewing someone with it. Before we get into it, I'd like to point out that when starting off, you want people who can potentially market you as well. If you're starting an investing podcast and by some miracle manage to land Warren Buffett, he's not going to do much to promote you because he doesn't do social media and the like, nor does he talk about anything other than his own companies.

So screen your guests in this manner. Remember, your first few months are all about growth.

Databases

This is a paid option but it is a worthwhile investment. One of the best ways to get to know the movers and shakers of a particular industry is to join a media database. Here are the big ones that are out there currently (Gregory, 2017):

- Gorkana
- Vuelio
- Meltwater
- Agility PR
- Response source

By registering yourself as a journalist/content creator, you'll be able to receive emails from PR people and other marketing professionals in your niche. As I said before, this is a paid option so you might not want to start off this way.

Using Amazon

Amazon's bookstore is a great way to find people who are major influencers in a particular area. I've already mentioned the BSR, and you can use this to find relevant authors and experts in your field to interview in this manner. You can also navigate your way to Amazon's bestseller lists and then drop-down by category to find the bestselling books and authors.

The tricky bit is contacting them. Most of them will have a website with a contact me section in it, but sometimes they don't. What you can do in such instances is to head over to whois.com and paste the url of their homepage in there. This will give you the registered email address of the person who is the main contact for that website.

The best part about this method is that Amazon is full of mid-level experts who are looking to expand their footprint. They definitely have some marketing going on and are looking to be cited and interviewed. You can approach these guests directly without citing them in a blog post (by these guests, I mean the mid-tier ones not the top influencers).

Sometimes you might find that the author has no published contact information and has no website. What you can do in such instances is head over to Twitter or some other social media platform and message them there. Most influencers are on Twitter, so you should be able to connect with them there. I'll address your own social media channels in the next chapter with regard to setting it up for maximum growth and receiving positive responses.

HARO

HARO stands for Help a Reporter Out and is a service where reporters send queries for interviews with experts as well as questions that need answering. You can use HARO to boost your own authority as well as find good guests. In terms of using it to boost your own authority, the way HARO works is this.

You'll receive two emails daily with a bunch of requests from reporters searching for sources. You can respond to these requests and if the reporter likes what you have to say, they'll quote you in their article. Millions of people read this article and the next thing you know they're Googling you and your business, thus bringing you traffic. Google will recognize the backlink or the fact that you've been mentioned and your domain will receive a boost. Additionally, you get to display the media source's logo on your website forever, which only adds to your credibility.

Other Podcasts

Well, if you can't beat 'em, invite 'em! Your competitors need not be your sworn enemies. You can invite them onto your show and gain referral traffic from a source you know has interested listeners. You stand a better chance of landing a big name here if you're in the same niche as they are.

Inviting Guests

The next step is to invite your guests onto your show. This isn't exactly rocket science. Get in touch with them or their publicist and send them an email mentioning the focus of your show and your stats. Mind you, when sending requests to famous people you will need to list your own stats as social proof.

This is why I mentioned earlier that it's best to have some content on your blog first with hopefully at least one backlink to it. Ideally, you'll have a decent social media profile by now as well (I'll cover this in the next chapter) so including your URL and your social media profiles should bring you a positive response.

Give them as many time slots as possible to interview and be prepared for them to say yes.

Chapter 7: Free Traffic - Part Two

The previous chapter covered the intricacies of Google and then moved into landing interview candidates for your show. In this chapter, you're going to learn all about using social media platforms wisely to promote your show's organic reach as much as possible. Again, our approach remains the same: Your podcast is just one branch, even if it is a major one, of your entire content creation strategy.

Gaining organic attention from social media is pretty tough these days. This is because every platform except for YouTube, Twitter, and Instagram to a certain extent has figured out how to make ads work for them. This doesn't mean organic reach is dead. There are many things you can do to be noticed and to prove credibility to those who check you out.

Think of your social media presence as being your business card. These profiles are not so much for your audience as they are for your peers in the industry. The platforms highlighted here are addressed with this in mind. Right off the bat, let me just say Facebook is not worth it. It's great for paid ads, but don't waste your time in the beginning trying to create a perfect page for yourself. Your mom is the only one who's going to see it, and she's your fan anyway.

Facebook will come in handy once your show grows, and I'll address that in the chapters on monetization.

YouTube

YouTube (YT) is one of the best platforms right now for organic reach. The great thing about YT is that it is a part of

Google and many keywords searched in Google produce YT search results. This is great because the competition on YT is far less than what it is on Google.

You might notice that a blog post about a certain topic doesn't make it onto the first page but a YT video on it does thanks to being relatively popular on that platform. Google designates certain keywords as video search keywords, and you can unearth these during your keyword research process by simply typing it into the search bar and seeing whether any YT results show up.

As the world's largest video search engine, YT is pretty powerful by itself. So let's take a look at how you can take advantage of this.

Optimization

Since your videos are just your podcast shows, you won't have too much of a problem creating content. Simply upload the video onto YT and you're all done! Unlike with your blog, I don't recommend creating supporting content for your videos since your blog will do most of the heavy lifting in that regard.

What you should do is mention your blog and YT channel on your show and tell people to subscribe, leave a comment, sign up for your newsletter and whatever it is you have going on. Most beginners make the mistake of uploading their videos and leaving the title the same as what they named it on their podcast host. What I'm saying is, don't name your videos "S1E2: Mo' Money No Problems by Don Snow." As much as you love your name Don, YT doesn't know who the hell you are.

Instead, use your keyword research to find the relevant keyword for your topic and name your video using those search terms. You'll be mentioning your podcast's name during your intro and people will figure out that it is a podcast, so don't worry about them not knowing. You can also mention this in the description.

The video description is a great place to leverage on YT. This is where you can paste the transcript of your video along with the timestamps. Now, I'm not a fan of pasting the entire transcript here. The thing to do instead is to paste just a few relevant quotes that contain related keywords and direct them to your blog where the full transcript is.

Make sure you timestamp your videos by topic so your viewers will know at which point you start talking about a certain topic. When naming these topics after the timestamp, use keywords again. You can find related keywords by looking at the drop-down suggestions YT gives you.

Competition

While the competition on YT might be less, this doesn't mean it's nonexistent. In fact, you can use your competition to help you with finding relevant keywords. First, take a look at their titles and their descriptions. This is where all of their keywords will be present. You can even string together a series of shows by borrowing their topics and improving it for your listeners.

Next, YT allows you to tag your videos with a few keywords when you upload them. These are not necessarily keywords that are searched by the user but can be thought of as broad categories which helps YT categorize your video. In some niches, these are pretty obvious. In some, they're not.

The best way to figure these out is to click on the best performing video in the topic (the one with the largest number of views) and then view the source code of the page. You can do this pretty easily in your web browser by right clicking and selecting "view page source" or "inspect page source." This gives you the raw HTML under the hood. Press ctrl+F and search for the string 'keyword.' This will give you the keywords used on the backend for any YT video.

Growth

Your YT channel will grow in lockstep with your podcast and your website. There's nothing special you have to do beyond remaining active on YT and releasing/uploading new videos as much as possible. You can release your podcasts on YT on the same schedule as you do on your podcast host. Just hold back on your blog initially until you receive some traffic and then go all in over there.

You can consider promoting exclusive content on YT to boost it as a standalone media outlet for you. However, this requires a lot of time and you'll be spending a lot simply producing your show. So ration it wisely. YT growth is organic, and one of the best ways you can ensure your viewers remain on your channel and explore it more is to create a great channel page.

The channel page is where all of your videos will be listed. Create sensible playlists for people to watch and create a decent trailer video for your channel. This need not be a big video; you can edit a few clips from your other videos to give people a gist of what your channel is all about. Make sure you include references to your podcast and website in your 'About' section so people can explore your stuff further.

Again, you might find that people are willing to remain on YT instead of heading over and subscribing to your podcast. YT traffic is the same as blog traffic, so I wouldn't worry too much about this. Think of YT as being your videocast monetization channel and monetize it as I'll show you in a later chapter.

SRT Files

SRT files refer to the video transcript files. You'll already have this in hand, so make sure you upload it to YT. The algorithm parses the file to find more relevant keywords. This is why it isn't a good idea to paste the entire transcript in the description since it clutters that area and YT does it on the backend through the SRT file, anyway.

All in all YT is a great way to gain additional exposure, so make sure you use it well.

Twitter

Twitter is the best social media platform for organic reach. The reason for this is ironically due to the fact that it is perhaps the worst run company among the social media giants (Mckay, 2019). While the others have figured out how to monetize and integrate ads as a majority of their revenue source, Twitter hasn't quite done this as well. Even YouTube, where ads are a distraction and are inherently unsuited for the platform, has done a better job than Twitter.

Management incompetence aside, this is great news for you. What this means is that it is easy to network and leverage traffic for your podcast. You won't be using Twitter in the same manner as you would YouTube. While YT is primarily a channel for you to gain more viewers, Twitter can be thought of as being a way to network with other influencers who can send you traffic wholesale.

Here's how you do it. The first and most obvious step is to create a profile. Mention your business' name and choose an eye-catching profile picture. If your niche supports it, always post a professionally shot headshot. If you're in an informal niche, such as photography, posting a picture of yourself in a business suit is not very appropriate.

Your approach to Twitter must be the same as the one you adopt when you launch your blog. Post some helpful content, stay active, and then start networking as much as possible. Don't focus on posting too much original content in the beginning. It's not as if you can post content on here.

Make sure your cover photo highlights your podcast and get networking.

Networking

So how does one network on Twitter? Here's something you should understand. Influencers on Twitter get contacted on a daily basis. Almost every single DM they receive is someone asking to collaborate, with them or do them some sort of a favor. The blogging tactic of mentioning them by name and then asking them for a retweet also doesn't always work. So what should you do?

In a nutshell, provide value. Start interacting with their accounts and retweet or respond to them. See if they have some issues you can help them out with. Cast a wider net by searching for your keyword in the search bar. If someone has asked a question about your topic, help them out with a response and follow them. If you keep providing value to these people, you'll find that your follower count will grow and you'll also start receiving traffic to your podcast or blog, whichever you choose to highlight more.

Use the same tactic with influencers. You do have to be a little proactive with them. See if you can spot something on their blogs or profiles that indicates they have an issue and genuinely help them out. The more value you can add to their lives, the more likely it is that they'll follow you back. Once this happens, the floodgates open.

You can message them directly and highlight your blog posts where you mentioned them and so on. Ask them if they would like to add any additional information on what you wrote and let them know. Don't ask them to appear on your show immediately. I mean, it's called social media for a reason. Be social and be a person, not an internet marketer.

As long as you're active and are posting tweets daily and interacting with people on there, you'll find Twitter a great place to network and leverage that into free traffic for your podcast. Just go about it like a regular person and you'll be fine.

LinkedIn

LinkedIn is the tie and suit wearing professional in the world of social media companies. Its 'professional' reputation has a lot of people thinking it isn't a social media company as much as it is a resume creator, and this is unfortunate. When used well, LinkedIn works wonders for the niches it applies to.

Now, you're probably not going to source a lot of visitors to your podcast from here, but what you can do is network much like you would on Twitter. Unlike Twitter, LinkedIn is built for networking, so it is a whole lot easier and you have more powerful tools at your disposal.

Having said this, it's not as if it's going to work for every niche out there. Generally speaking, if you want to get in touch with people in professional niches or those who are high up the corporate ladder, LinkedIn works best. A good example of this is the law niche. Another example is the personal finance niche if you wish to interview a CFA or a CPA who can help with any accounting and so on.

LinkedIn is great for organic growth and networking. The key on this network is to stay active. Unlike the other platforms, its organic reach is pretty high despite having a robust ad system. So let's look at how you can increase your organic reach on here.

Profile Basics

On LinkedIn, the longer your profile and job history is, the better. The algorithm favors profiles which are completely filled out and have literally every box filled with some information. As a podcast marketer, you need to tick all of these boxes. LinkedIn also places a heavy emphasis on keywords used throughout your profile.

The way to do this in your profile is to make it sound as natural as possible. One of your first few lines should involve the words "host of xyx podcast." It's important to get this out right off the bat. When people view your profile on mobile or on the desktop, they'll see the first four lines of your profile description, so make sure you advertise your podcast over there.

Do not create a job title named "host of xyz podcast" since that doesn't come across well when someone views your profile. You want people to think you do more than sit in front of a mic talking all day. Instead, you can give yourself a title such as "founder xyzblog.com" and so on. Pick CEO if you want. Remember, your podcast is just a cog in your entire content machine even if it is your primary focus.

Your next area of concern is your title. On LinkedIn you'll notice there is a short space to describe what you do next to your name. There's two schools of thought to filling this out. The first is to have short and sweet titles, and the other is to have a short sentence that describes the value you provide. Which approach you take depends on the goal you have for LinkedIn.

Personally, unless you happen to be a well-known public figure, it's better to have titles next to your name instead of a sentence. The great thing about LinkedIn is that you don't need to restrict yourself to just one title in this space. What most people do is use the "title 1 | title 2 | title 3" format.

These words serve as both an introduction to what you do as well as keywords when people search for a particular expertise. Don't overthink this part and simply describe the things you do as part of your niche.

Groups and Updates

LinkedIn groups are a great way to network and put your name out there as an expert of whichever topic you're aiming to talk about. There's no secret sauce to getting noticed in groups. Simply post regular updates and like posts which seem relevant to you. One powerful method of leveraging LinkedIn is to expand your list of connections beyond 500.

This may sound daunting, but it is actually very easy. Most people are very open to networking on the platform and you'll find that a simple request is enough for them to accept you as a part of their network. Once you cross 500, LinkedIn displays the number as "500+" and this works as social proof when people look you up.

When you publish updates to your blog, make sure you publish them on LinkedIn as well or highlight your new blog post as a post on LinkedIn. People will take note of it and you'll find people clicking onto your website, and some will even check out your podcast. However, the point of LinkedIn is to network with possible guests for your show. So how do you do this?

You could send them a connection request and if they accept, you can introduce yourself and thank them for connecting with you. Don't ask them to be a guest immediately since it's a bit too soon for that. Instead, engage with their posts and leave comments. They'll notice your updates as well and eventually you can get around to asking them to be guests on your show.

LinkedIn does have a premium feature where you'll be able to send requests to people who are beyond your second degree of connections. A second-degree connection is a friend of a friend. You can message this person a connection request, but someone who is completely outside of your network can usually not be messaged. This is where LinkedIn's premium InMail feature comes in handy.

Now, if you were looking to land some business leads, this is a good idea, but frankly, for a podcast, it doesn't have much value. Just remember that beyond a certain point of searching and connecting with people, LinkedIn will prompt you to sign up for the premium feature. So take it slow with this.

As I said before, there isn't some secret sauce to the way LinkedIn works. You just remain active and keep expanding your network and post updates. Connect with people you would like to interview and slowly build your network.

Apple Podcasts

Formerly a part of iTunes, Apple has since broken this out into a separate app. The podcasts app in iOS has a bunch of features and also has one of the largest user bases for podcasts in the world. If you manage to focus just on Apple and did nothing else, you'll still do extremely well.

Of course, everyone else has this same idea and as a result competition is very high. This is why it pays to have a well-rounded content strategy in place. That way, you can diversify your traffic sources. I've already mentioned one of the keys to being found on iTunes (I'll call it this for simplicity's sake) is to be featured in either the 'hot' category or 'new and noteworthy' category.

Before looking at these, though, let's look at another iTunes feature that is extremely important for you.

Rankings

Within the app, iTunes has a number of categories and subcategories. If you click on any of these you'll be shown the top-ranked podcasts in them. A high ranking is a self-fulfilling prophecy of sorts since these shows are displayed the most prominently and a user will naturally click on them. These rankings are determined based on the number of subscribers the podcast gets over a period of two to three weeks (Mclean, 2018).

Note that I said subscribers. Not downloads, likes, or comments but subscribers. Therefore, wherever you advertise your podcast, make sure to prompt people to subscribe. The greater the number of people who subscribe to your podcast, the higher your rankings are going to be.

This is why it's important to have a good backlink and influencer relationship in place before you drop your first show. To do that, you need good social media profiles and a few value-adding blog posts. My point is that to prepare a podcast, you need to start well in advance to be able to rise up the rankings as quickly as possible. The quicker you can get a high-profile guest onto your show, the greater the number of subscribers you'll receive and the more iTunes will push your show.

This sort of advertising is invaluable for your growth.

Categories

The "new and noteworthy" category has been talked about already. The thing with this category is that almost everyone will get featured here. The key to using this category well is to keep releasing new content regularly and as much as possible. I've already mentioned how daily releases are a great idea to be featured on here continually.

The other category is the 'hot' category that highlights the podcasts with the greatest increase in download and listen count (not subscribers). There's no way to hack your way into this category, I'm afraid. Well, there are black hat techniques but I strongly recommend you don't do these. Apple will catch you, and you'll be banned for life.

So the takeaway from all of this for you is to schedule your podcast release thoroughly and lineup all of the various factors prior to launching. Buying a mic and then uploading it onto iTunes is not going to get you anywhere. For exponential growth right from the start, you need to build backlinks, network with influencers, and have them push your show. To do that you need to provide them value and display that you're serious about your niche.

Email Marketing

Email marketing is the oldest of the old school marketing methods and is still the most powerful method of bringing eyeballs to your content (Mclean, 2018). Email marketing is quite simple to practice. You simply need to run a newsletter to all of your subscribers. The key question is how do you get subscribers in the first place?

Well, this is where your blog comes into play, and it is why you need to publicize it on your social media platforms and marketing avenues. The best and perhaps the only way for you to collect emails is by offering visitors an incentive to sign up. In the internet marketing world this is referred to as a lead magnet. As the name suggests, this is a free item (a magnet) you offer in exchange for an email (a lead).

Let's look at this in some more detail.

Lead Magnets and Newsletters

The most common form of lead magnet blog owners employ is a free eBook that gets delivered to the customer's email address. You could write a 10,000-15,000-word eBook on any topic in your niche and offer this as a lead magnet. If you feel this is too much work, you could outsource it to a ghostwriting company like The Urban Writers who will do the job for you.

Next, you set up an account on a software such as Mailchimp and upload your lead magnet and the email text you will use when you deliver it to your customer. On your website, you can install a simple plugin (if you're using WordPress) to collect the email and integrate it with Mailchimp on the backend. The email shoots off all by itself and you have a valuable email address.

Emails are valuable not just for free ads but also for paid ads as you'll see in the next chapter. Either way, you can design a great newsletter template using free software like Canva and start sending your customers news about your podcast. In addition to this you can use newsletters to monetize your content machine in other ways as we'll shortly see.

Email Marketing Tips

One of the best ways to get people to open your emails is to track your open rates and use what is called the double-open strategy. This is how it works. You send the first email, having crafted the perfect subject line and text along with the relevant Call to Action buttons (CTA). You'll notice that your customers won't open the first email you send.

The way to get these people to open or to test whether they're genuine customers for you is to send them the same email but with a different subject line. You'll have to try a different line and make it more captivating, but generally speaking, this will increase your open rates.

Speaking of subject lines, one of the best methods to get people to open your email is to use the cliffhanger technique by leaving an ellipsis at the end of it. This is how your local news channel has been advertising itself for years. "A deadly disease in your neighborhood? Tonight at 9…" The email subject line equivalent would be "This is gonna be huge…" or something of that sort. Don't overdo it, though, since your customers will get wise to it.

Most people use Gmail, and this makes another section of your email extremely important. Gmail displays a preview of the first few words of your email so you need to make this as captivating as your subject line. Therefore, do not insert a header into this space since it will make no sense in the preview, even if it looks pretty on your template.

The text of your CTA buttons also needs to be more than just "click here" or "shop now." The best way to get people to click on them is to craft contextual buttons. So instead of saying "listen now," you could say "save 50% more money this month" (assuming you're hosting a personal finance podcast).

When it comes to monetization, you want to play the long game with your customers. Yes, money right now is great, but you don't want to turn off a potential lifetime customer, do you? I'm using customer and listener interchangeably here by the way. After all, your listeners are really your customers and drive your monetization.

Some marketers rush to push affiliate products to their email lists, but this is a bad thing to do. Imagine giving someone some personal information and the next thing you know they turn around and start hounding you to buy stuff. In real life, salesmen never do this, but somehow people think it will work on the internet. Like your blog audience, take the time to build a relationship.

Carry regular maintenance on your email list. Have an email that you can send to people who never open your emails and ask them if they wish to unsubscribe. If they don't open even that email, remove them from your list. How many emails should you send before this one? Well, eight is a good number. If your frequency is once every week then that's two months you've given someone, which is a long time.

Lastly, keep your emails short. The reason is that Gmail diverts longer emails to the 'social' tab and your subscribers will never see it in the 'primary' tab. Graphics heavy emails also land in the alternate tabs, so despite your desire to create a pretty newsletter, keep it short and simple.

Chapter 8: Paid Traffic

It's no secret that podcast discovery is a huge issue in this business. Everyone has a problem with it, and the apps that host/deliver podcasts are constantly fiddling with their algorithms to deliver the most interesting content to their listeners. Still, many users complain about how hard it is to find interesting podcasts.

Paid advertising is one of the best ways to get yourself noticed. Free traffic takes time to build and will sustain you over the long-term. However, in the short-term, there's no reason for you to keep talking to yourself. While there are many paid options out there, I'm going to concentrate on the three best avenues for you to utilize.

Facebook

Despite being one of the most maligned companies in the world, Facebook is still a massive outlet for advertisers (Gibson, 2018). Thanks to its captive user base, which is the largest in the world, and its robust targeting options, Facebook (FB) ads has the potential to deliver massive ROI on your ad spend. Well, that's the good news. It's best that you hear the bad news as well before proceeding.

Facebook is not exactly a trustworthy company (Gibson, 2018). Their reputation is well deserved after their recent shenanigans with selling their users' data. This less than ethical behavior transfers over to their ad platform as well. When FB Ads started out many years ago, the platform was pretty straightforward and easy to use. These days, though, it is a far more complex beast and is, in my opinion, not suitable for beginners who don't have too much money to spend.

Money is at the heart of how this platform behaves. At the end of the day, FB makes money on the amount of ad spend you allocate. This doesn't mean they'll give you useless clicks. However, in the initial stages of your campaign, you'll find that you'll end up spending a lot more than you expected with your metrics all over the place.

In my opinion, the best way to use FB Ads is once you have some data in place that you can give the algorithm. This way, you'll avoid that initial audience discovery period that FB undertakes. Hence, if you're at the beginning of anything, be it your audience discovery process or in terms of FB ads experience, you want to avoid this platform. If you do have some data about your audience, let's look at how you can leverage Facebook.

Lookalike Audiences

By audience data, I mean hard facts about their demographics and behavior, not what you think your audience looks like (your listener persona). The question is, how can you determine this? There's only one way, and that's by email. Your customers' emails have a wealth of information about their behavior and while you can't utilize it, Facebook can.

Your first step on FB should be to set up a page for your show. You can't run ads without this. Some hosts opt to use FB as a forum for their show, and you can do this as well once you're up and running. For now, don't worry about getting likes for your show or anything like that. Populate it with a few posts from your blog and then head over to the Ads Manager.

Once in there, what you want to do at the Ad Set level is upload your email list and create a lookalike audience. This means you're telling Facebook to create an audience whose behavior mimics that of the people on your email list. You will need to select a few more options, but these are pretty easy to navigate and figure out. For example, you can restrict the countries you wish to advertise to and so on.

Here's why this process is so powerful. Your original email list has people who have already qualified themselves as being interested in your show. Instead of starting from scratch and telling FB to find people who might be interested in your show based on what it's about, you're now telling FB to look at just a subset of its user base to determine who might like your show. As you can imagine, the hit rate is going to be a lot higher.

In terms of metrics, FB throws a lot at you, but the most relevant one you want to look at is your CPA. You can set up campaigns to redirect people to your show page. However, a much better way of achieving this is to keep your customers on FB itself. You see, FB prefers it if people remain on their platform. This doesn't mean they'll hobble ads that remove people off FB, but there's a clear conflict of interest here so why risk it?

What you need to do is use a plugin that will post your podcast feed directly onto your Facebook page. This way, people can then click on the link in your post and they can listen to it either on the hosting page or on FB itself. Even better, if you upload the podcast as a video on your podcast host, you can post it as a video on your page. This means people can view your content right on FB, and everyone's happy.

I've said this twice already, but I'll say it again: The worst way for you to use FB ads is to go in without any idea or data about who your audience is. This will only result in useless ad spend and you'll end up spending way too much before FB can figure out who is interested in your show. So collect emails first and then create lookalike audiences to speed up the process.

Google Ads

We've looked at Google for SEO and free traffic and that was quite a mouthful to swallow. Don't worry, I'm not going to write a novel on Google Ads. This is mainly because Google Ads are actually quite straightforward when compared to SEO. They work really well and much better than FB ads when it comes to producing results, in my opinion. The only downside with them is that if you think FB's analytics can give you a headache, well, Google takes it to another level entirely.

There are tons of options within Google Ads with regard to how you can optimize your campaign, but experiencing success with this is quite simple. I mean, you can complicate it if you want, but for your purposes, there's no need to do so. Let's take a look at this.

UX and UI

Google Ads is governed by the same principles as SEO, largely speaking. What I mean is things like the bounce rate and relevance are extremely important. Google Ads revolves around a thing called a Q Score, which stands for quality score. A few things go into this. First, how relevant are your keywords to your ad and topic? Are you advertising kayaks and using surf boards as a keyword?

Next, Google looks at how well your landing page matches up with your ad text. Are you advertising something else entirely? Lastly, it looks at the bounce rate. If a person clicks on your ad and then immediately navigates back to Google, this is a blaring sign that your page isn't very relevant to that ad. Your Q score also impacts the amount you pay per click. Thanks to this, you can end up paying a lot less than your competitors and still outrank them (unlike FB which doesn't have the competition model and revolves around money spent).

All of this leads us right back to our old friends UX and UI. Your first task is to make sure this is top-notch. What a lot of podcast hosts do is they redirect people right to their show's page. Think about it: Someone has no idea who you are and has clicked on your ad looking for a solution and instead finds they need to listen to an audio podcast in order to get this solution. What are they most likely to do?

This is why having different pillars of content is so crucial. You should not redirect your users to the show page. Instead, direct them to a helpful article that contains references to your show. The ideal scenario would be them clicking onto your helpful article (which can be a show transcript) and then clicking another link on that page which takes them to a page where they can subscribe or enter their email address.

Don't make show subscription your primary aim with Google Ads. Instead, capture their emails and then pitch your show via newsletter or through other media channels. Embed your video podcasts in your blog posts so they can find you on YouTube and subscribe to that channel. Don't restrict yourself to just the audio method since this makes no sense and actually limits your ability to monetize your traffic.

Overcast

Overcast is one of the podcast listening apps I alluded to earlier in this book. Well, did you know that you can advertise on it? While its ad algorithm is nowhere near as sophisticated as the previous two, Overcast can deliver huge ROI for you because you know you're advertising to people who are definitely interested in listening to podcasts. I mean, why else would they be on Overcast?

Setting up and running ads is simple, but on the flip side this simplicity means there isn't much you can do to tweak your ads beyond a few cursory changes. Traditional PPC (Pay Per Click) ads require you to run split tests and do all kinds of tweaking, but this isn't possible here. This can be both a good and a bad thing.

Experienced PPC ad managers will find the experience frustrating, but beginners will find it pretty easy to navigate. One of the downsides of Overcast, which affects everyone, is that the effectiveness of their ads is decreasing thanks to more podcasters jumping on board. But first, let's take a look at how it works.

Slots and Spend

Overcast's platform works in the following way: As a podcast owner you get to buy slots for different categories. These categories are the ones that are most relevant for your show. As of this writing, Overcast has the following categories ("Advertise Your Podcast," 2019):

- All
- Arts
- Business
- Comedy
- Education
- Fiction
- Health and Fitness
- History
- Kids and Family

- Leisure
- Music
- News
- Religion and Spirituality
- Science
- Society and Culture
- Sports
- Technology
- True Crime
- TV and film

Each category has a fixed number of advertising slots, and each slot is valid for a month. You pay a fixed price for a slot, and that's it. The prices of slots vary based on demand. For example, you might find Business selling for $1,150 per slot, but Kids and Family sells for $90. Slots have started selling out far quicker these days as more podcasters are migrating to Overcast to make use of the advertising option.

The ad itself isn't anything fancy. It's a simple banner with text that shows up below the podcast player from time to time on a user's screen. You can advertise in multiple slots, but this will ratchet up your spend quickly. Thankfully, Overcast provides approximate metrics to measure how you can expect your ad to perform.

For example, a slot in Business costs $1,150 as of this writing. Your ad runs for thirty days once you pay. Overcast estimates you can get between 2,000-5,000 taps in this time, and this will convert to 150-200 subscribers. So you'll pay around $7.60 per subscriber. Are you making that much through your monetization channels? We'll deal with this in the next chapter.

To be honest, paying this much per customer is a bit ridiculous. A manageable spend per customer would be somewhere around $2-3. Another thing to be careful of with Overcast is that the metrics will change rapidly overnight. This is due to the fact that a user who taps your ad is not going to subscribe immediately. They'll listen to a few shows and then subscribe. So you need to give the analytics some time to marinate.

In terms of analytics, taps and subscribers along with spend is all you get. In fact, if Overcast feels you need not spend the entire amount gaining subscribers, they will refund a portion of your money back to you ("Advertise Your Podcast," 2019). This behavior will not last long, given the track record of tech companies, so you might as well make hay while it lasts.

Other Methods

There are some other avenues you can use to advertise your podcast. A good option, which is often ignored, is the use of billboards. Yes, I'm talking about physical billboards. You can use a service like blip, which will handle all of it for you. This is old-school advertising, and it's not going to work for every niche. However, if you're in the health niche or religion niche, you can see great results. This is doubly true if you operate in a smaller city or town.

A good option to utilize is to purchase banner ads on podcast search websites such as Listen Notes. This is the biggest podcast search engine and allows people to curate their playlists. You know podcast listeners are on the site, so you might as well advertise on it. If you're running Google Ads and choose to run ads on the display network, your ad will show up on Listen Notes as well.

Lastly, if you're in a specific niche that interests a set of people such as lawyers or students and such, go to places where they congregate and start handing out business cards. This is just the real life equivalent of LinkedIn networking. Do whatever it takes to spread the word.

There's no single method that is going to bring you rip-roaring success. You need to experiment with various options and see what works for you best.

Chapter 9: Monetization

How much money would you like to make from your podcast? How much do you think you can make? Well, to answer this question, we can take a look at the data that has been compiled by AdvertiseCast, which is an ad network (more on this shortly). According to them, an average podcast show makes somewhere between $0 to $100,000. If there's a prize for useless statistics, this has to be in the running for top spot.

Despite that stat giving us nothing useful, it clearly shows the potential that is inherent in podcasting. The truth is that there are many ways for you to monetize, and the higher-earning podcasts routinely diversify both their income streams as well as their content channels. Hopefully you can now understand why I've been going on about having multiple content channels.

By having multiple channels you're opening yourself up to pretty much every online monetization method there is. Not to mention the fact that you'll be able to attract more listeners. As you'll see now, the number of subscribers you have is central to how much you can get paid.

Sponsorship

This is the easiest and preferred way for most podcasters. The catch is that no advertiser is going to pay you unless you have a large enough subscriber base. If your show has two commercial breaks, the money you earn from sponsorship does add up. Typically, a thirty-minute show will have to have breaks. One right before the show begins and one in the middle, around the fifteen-minute mark.

Pat Flynn's AskPat podcast earned around $3,500 in December 2017, and according to Flynn, he has earned over $350,000 since starting his podcast in 2014. Of course, he has other monetization methods that increase his monthly income, but this is purely from advertising.

While figures like this might seem great, you need to keep the audience experience in mind. Littering your podcast with ads might increase your revenue in the short-term, but you're not going to hang onto your subscriber base for too long by doing this. Let's now look at how podcast advertising payouts work.

Payouts

Podcast ad sponsors pay you on a CPM basis. For those of you who are unaware, this stands for Cost Per Mille. A mille is 1,000 impressions, and an impression is a single play of the ad. So, if you mention your sponsor and run their ad once and if you have one thousand subscribers, you've garnered 1,000 impressions of one mille.

According to Advertise Cast, the average rate card is $18 CPM for a 30-second ad and $25 CPM for a 60-second ad. Larger podcasters can command higher than average rates of $25-$40 per 30 seconds. So, if they run a minute's worth of ads on their shows and if they have a million subscribers, well, you do the math.

While it is possible for you to earn 100% of that sweet ad income, for purposes of sanity, I don't recommend handling sponsors by yourself. The reason is that you have your hands full already creating content. This is your primary job and you're better off leaving this to a company. This brings me to the role of ad networks.

How to Find Sponsors

A podcast ad network is a marketplace where you list your podcast and the network matches you with sponsors based on your statistics. A popular network is the aforementioned Advertise Cast, but there are many others. Boardwalk Audio is another example used by Alan Johnson, who runs a podcast about comedy writing.

Here are some other networks for you to consider:

- Midroll
- Archer Avenue
- Podgrid
- PodcastOne
- Megaphone
- Authentic Shows

The network will usually take a 20% cut of your ad revenue. In my opinion, this is worth it since it saves you a lot of time and hassle. Also, by appearing on a network, you've been validated to a certain extent in the eyes of a sponsor, so you'll stand a better chance of landing sponsorship.

If you're adamant about going at it alone and still cannot find sponsors, then consider offering free airtime to sponsors. Let them get used to the increased traffic and then charge them according to the standards mentioned previously.

Crowdfunding and Donations

Crowdfunding is a great way to tap into the loyalty of your listener base. One great example of this is the Embedded podcast hosted by Elecia and Christopher White, who crowdfunded a new set of microphones to use to interview guests (Fang, 2019). They used Patreon specifically, but other podcasters have used sites like GoFundMe and other crowdfunding sources.

It might hurt your pride to seek donations like this, but it really depends on your topic. If you're starting a podcast to highlight a real issue and are extremely passionate about it, then it is worthwhile to ask for your listener base to spend some money and donate to your cause.

Selling Services

Your podcast is the best place to pitch your services and products to people. This is a pretty straightforward way to monetize your show, to be honest. The only restriction is whether there is a service you can offer your listeners. The most obvious product to offer is your time and attention. In other words, a consulting or mentorship service.

If you're in the making money or business space offering, this is a no brainer. Even if you are a business, offering a discount to listeners can be a great way to attract new leads for your services and products. However, what are you supposed to do if you find that there are no services you have to offer? Well, that's when you get creative.

Masterminds and Escapes

The one thing above all else you should be offering your listeners is a sense of community. You can call this branding or whatever you like, but ultimately a listener remains a loyal one thanks to the sense of community you build and the value you provide. If you wish to monetize over the long run, I would say that you need to put your sense of community above everything else, even monetization.

This is because you can always figure out ways to monetize your show as long as you have a good audience. By good, I mean a loyal and committed audience that is willing to take action on your direction. This makes it sound as if you're starting a cult and not a podcast, but you get my point.

A great way to foster this sense of community is to hold a meeting where listeners can converge and interact with one another. You can call it a mastermind or a community escape or whatever you like. In fact, you can create a few episodes around this event and entice more people to join.

There are shows that pitch mastermind events for a fee to their listeners. Masterminds are usually associated with business and entrepreneurial shows and are usually held at exotic locations close to coworking hubs such as Bali, Thailand, Lisbon, etc. You don't need anything that fancy even if you aren't in the same niche. Giving your listeners the chance to interact with you and one another is a great thing to do.

So what can you charge for such events? Well, it's entirely up to you and the value you provide your attendees. This is the great thing about having a loyal listener network. Nick Loper, the creator of Side Hustle Nation, charges $97 for access to a private mastermind group (Fang, 2019). Mind you, this is just online access. Within this group, he pitches additional services of his such as mentorship and in-person masterminds that cost more.

You can even pitch your services as a speaker on your topic to your listeners. Don't come right out and say "hire me as a speaker," but you can mention it in a creative way. If you come across as an expert or as someone who can offer people value, you'll find that your listeners will naturally start messaging you asking you whether you're available to speak somewhere.

Memberships

Memberships are a great way to leverage community loyalty. The most obvious and boring way podcasters go about leveraging memberships is by charging people a fee to listen to their show. This is a bad idea since it limits the number of people who will sign up to your show. Growing your user base should be of paramount importance to you, whether you have ten subscribers or a million. Do not ever charge people to listen to your current episodes.

One method that podcasters implement is to copy what Dan Carlin, the host of Hardcore History, does and charge listeners a fee to listen to older episodes while providing the latest ones for free (Fang, 2019). I would suggest not doing this either since a listener who likes your current show and wants to listen to you loyally will go back to the first episode to start afresh. Charging them a fee for this is a slap in their face. So I'm not particularly fond of this and would consider it a last resort of monetization.

You need to be more creative than this. One method is to offer additional goodies for a charge on your website. Have a section that is for members only and have these people pay a small fee per month for access. Within this place, you can offer additional stuff that free listeners won't get. What you wish to offer depends on your niche, of course.

For example, if you're in the entrepreneurial niche, you can create a premium area where you go deepcr into specific detail about a particular way of making money online and provide a step-by-step guide and the best method to implement a business model. In your free show, which is effectively an advertisement for your premium area, you can talk about things on a higher level while still providing value.

Stock investing podcasts do this quite often. The hosts talk about general sectors but don't mention any picks outright. The paid section of their websites is where people can sign up to recommendation newsletters with more in-depth research.

Another method of monetizing your old content is to provide the MP3s for sale. Sure, they can browse through your show's archives and listen to it for free, but they'll need to load and download the episodes. If they wish to binge listen to your show, they can simply buy the MP3s for a flat fee. You can set this monetization method up right off the bat by running series on topics asymmetrically.

For example, if you're in the make money niche, you can run parallel series on making money via podcasting, making money from a blog, monetizing social media, and so on. Run one episode on a topic and mix them up. This makes it easier to sell these audio files as a series down the road.

Recording Tickets

This is something you can do once your listener base gets bigger. You can partner with a venue and sell tickets to attend your show live. You don't need a fancy studio to record live. In fact, you can partner with a bar or a restaurant that is close to where you live and pitch this to them. You can give them a cut of the sales and offer discounted food and drinks to your listeners as part of the price of the ticket.

Currently, it is the big podcasts such as Pod Save America who do this, but as I mentioned earlier, it's all about how loyal your listener base is. While you cannot do this with a smaller base, you can conduct a road trip of sorts if you have a medium-sized base that knows you're keen on meeting them and interacting with them. Why not seize this chance to interview your listeners? This will make them even more willing to pay to see you in person.

Partnering with a local establishment has other advantages. For one thing, you can offer them free sponsorship and publicity in exchange for recording space. This is precisely what the podcast Mind Gap does with a bar named Elephant and Castle in Chicago (Fang, 2019). The bar gets free promotion and the host presumably drinks for free (to an extent). This also reduces your costs if you have space constraints or live in matchbox.

Reducing or minimizing costs doesn't get enough attention from podcasters but is actually a great way to boost your profits. I mentioned this back in the technical section as well. You want to minimize your costs to the point where you can still put out great quality audio and video. So it's not about just reducing costs. You can record video on a potato and use your laptop's mic, but that's not going to provide any value, is it?

Using your smartphone's mic and camera is a good example of minimizing costs. I would always recommend an external mic, but if you can't afford this, a smartphone is a great option.

Affiliate Marketing

Affiliate marketing is one of the best ways to monetize your online venture, let alone a podcast. In case you're unsure of what this is, think of it in this way: Let's say you're a salesperson working for a corporation. This corporation pays you a commission on the basis of how much you can sell. This is what affiliate marketing is all about. You're pushing someone else's product and earn a commission on a purchase.

In this scenario, you're the affiliate, the company whose products you're selling is the merchant, and the intermediary, which is where affiliates and merchants come together to get matched, is the affiliate network.

Affiliate Networks

There are standalone affiliate programs that merchants offer, but finding these one by one is going to take a lot of time. It's far easier to login to a popular affiliate network and promote one of those products. A popular affiliate program among podcasters is Squarespace, which both sponsors and offers affiliate programs to hosts. You can promote their service on your website using affiliate links and have them on as a sponsor for your show.

An affiliate link is a unique link that is provided to you by the affiliate network and lets the merchant know that you are the referrer. The best affiliate networks to join are:

- Amazon
- Clickbank
- CJ Affiliate

Between these three, you'll have more than enough products to promote. A lot of podcasters and online marketers sign up for Amazon thanks to the vast number of products the website has. Amazon usually pays you 3% on most products and offers a 24-hour period where you can earn commissions on anything else your listener buys from Amazon. So if they buy a yoga mat for $30 and a chandelier for $1,000, you earn commissions on both, despite recommending just the Yoga mat.

Clickbank is devoted to digital products and courses as opposed to physical products. To be honest, the quality of products on Clickbank is usually low. You'll find a lot of herbal tea weight loss types of programs here, if you know what I mean. In contrast, CJ is a far better network, but the merchants on there demand a certain standard and details about how you're going to promote their products.

If you're interested in promoting just online and digital products where the commissions are very high, you can check out JVZoo, which is dedicated to products in the internet marketing space. Given the niche there are many charlatans on here, so you want to be careful of promoting stuff. Research the product thoroughly and only then recommend it to your listeners and content consumers.

There really is no limit to what you can promote via affiliate marketing. A popular category of product that is being promoted is books and audiobooks on Amazon. You cannot provide a direct link on the audio version, but you can leave a link to it on your blog or in the description of your YouTube video.

Apparel and other merchandise are also great things to promote. In fact, you can promote the gear you use to record your podcast and also the host of your podcast. These companies have lucrative affiliate programs, so take advantage of them! Don't go overboard with it, however. Remember that your purpose is to provide great content to your listeners.

Merchandise

This is something Dan Carlin, who I mentioned before, does quite a lot (Fang, 2019). He hosts a shop on his website where people can buy gear that is related to the stuff he talks about on his show. If "hosting a shop" sounds like a lot of work, you really don't need to worry about this. The rise of print-on-demand (POD) services has made this easier than ever.

Websites such as Teespring make it really easy for you to upload designs and sell merchandise. You don't need to design the merch yourself, either. You can hire a freelancer on Fiverr and have them send it to you for a total cost of around $10. You upload this design to Teespring and can sell it on T-shirts, hoodies, mugs, posters, rugs… whatever you can think of.

Teespring will take a cut of your sales, but they do all the heavy lifting for you. All you need to do is simply upload the design and that's it. They create, ship, pack and handle returns on your behalf. It's extremely simple and easy.

While POD is a great way to get into the merchandise category, one particular type of product that is often ignored by podcasters is the info product.

Information Products

Information or info products are a real goldmine if you have a good listener base. An info product, as the name suggests, provides information about some topic. This can be a book or a course. Generally speaking, people are far more willing to buy books than courses, or at least this is what the common notion is. I feel this is wrong because this sort of thinking comes about as a result of incorrect pricing strategies.

You want to price your book as reasonably as possible and make it affordable for your listeners. Common pricing points are $9.99 for an eBook, $19.99 for a paperback, and you don't need to worry about your audiobook's price since this will be fixed by the platform you sell it on. Usually, this is Amazon's Kindle Direct Publishing (KDP) platform.

KDP makes it easy for you to upload your eBook and have it converted into a paperback. There are freelancers on Fiverr who will design your book's cover and convert it into a paperback format for you to upload it onto KDP. Once this is done, all you need to do is announce the release of your book and make it available on ACX, which is Amazon's audiobook platform. Over here, you can have your book narrated and provide listeners with codes they can use to redeem your book for free as a promotion.

Your books will bring you a steady income, but what you really want to do is parlay this into a course or mentorship of some kind. If this doesn't make sense for your niche, brainstorm some other way you can address the problems the book tries to solve but pitch it as a personalized service that you can charge for a lot more. I'm talking about pricing it at $3,000 or so.

If people buy the book, they have a problem. If the book doesn't solve their issues, signing up for personalized service makes sense since they obviously are not able to figure things out. In such a scenario, it makes sense to charge a lot more and they'll be willing to pay. A lot of podcasters will look at the $3,000 and think there's no way anyone would pay that. Well, that's because they haven't set things up the right way with the preliminary info product.

So setup two tiers of info products, and you'll have no problems monetizing your podcasts.

Virtual Summits and JVs

A virtual summit is a business in and of itself, and to be honest, and I could write a book as long as this one talk about this. As a result, I'm not going to go into great detail about setting up a virtual summit and the logistics of it since it will distract us from our main topic. If you wish to learn more about this method of making money, I recommend visiting baileyrichert.com and studying everything she recommends.

A virtual summit is exactly what it sounds like. It is a collection of speakers talking about a subject, with you interviewing them, and posting these interviews online. You can record the interviews beforehand and post them during the summit week. A summit usually lasts around 4-5 days, and you'll be releasing content (interviews) with these people during the specified time.

In return for access to this content, you can charge people with a one-time access pass. People are usually willing to pay anywhere from $50 to $500 for an all-access pass, depending on the quality of people being interviewed and the topics discussed. You can host the interviews on your own site, but it's better to use a service like Clickfunnels to create landing pages.

The best part about virtual summits is that you can use them as both a marketing tool as well as a monetization strategy. Think about it: No one might know you, but they will know the influencer you're interviewing. If you introduce yourself as the host of XYZ show and mention your blog, people will head over to those places more often than not, and if they like the sort of questions you're asking your guests, they will subscribe to you.

Just to make it clear, you will not be posting these interviews for free on your podcast show. Instead, once the summit is over, you can sell these for a price, often equal to the price of the all-access pass. Another reason why virtual summits work well to increase your subscriber base is due to the fact that the influencers are going to promote it to their email lists, which will be considerably larger than yours. Imagine five influencers with a million people in their audience sending traffic your way! Best of all, you're making money while promoting yourself.

A JV is a joint venture, and I mentioned this tactic earlier when talking about SEO. You could start a new project or create a new product with another influencer and charge people for it. In essence, this is the same as creating an info product, but this time the collaboration with the influencer brings more people to you.

Conclusion

As you can see there are many different ways and means of monetizing and marketing your podcasts. What's more, you'll find that there is no cookie-cutter path to podcast success. A good example of this is the case of Jamie Masters.

Masters is the host of the wildly successful podcast Eventual Millionaire and has interviewed over 250 multimillionaires on her show. By some accounts she is a millionaire herself thanks to her show. She's been featured in a number of national publications and is one of the early movers in the podcast space.

It wasn't always like this, though. At the age of 22, Masters found herself saddled with debt to the tune of $70,000 thanks to student loans and a job she hated and sucked the soul out of her. She was obsessed with figuring out how to become a millionaire and a lightbulb went off in her head. She would interview millionaires and learn for herself!

Masters didn't worry too much about monetization at this early stage. She was focused on providing value and increasing her listener base, which goes back to providing value. Fast forward to today, and Masters is a business coach and has a number of products she advertises on her website. She sells merchandise, her show has sponsors, she sells a mastermind program, and a course as well as a hiring kit to help entrepreneurs determine the right sort of employee they ought to hire. All of this is over and above the regular consulting she does.

Did Masters envision these many channels of monetization when she began her podcast? Well if she did, Nostradamus is going to have some competition. This is how a lot of podcasts work, to be honest. The hosts begin with a vision in mind about what they would like to do and they figure it out along the way.

You picked this book up in the hopes of trying to figure out everything you could about podcasting. Well, you now know a lot about podcasting, but there's still one element you need to figure out, and that is yourself. Making money is great, but you want to do it on your terms and while continuing to provide value to people. If you don't do this, you'll find your listener base eroding and income decreasing.

Keep this at the forefront of your venture at all times. Another part of figuring yourself out is to determine what style of show you want to produce and how you're going to monetize it. Different people have different methods of monetization that they're comfortable with. Some hosts lean heavily on YouTube while others focus heavily on monetizing and tying their blog in with their podcast, like Pat Flynn does.

So do what fits you best since this is what will allow you to shine. Remember, people are signing up to listen to you, first and foremost. Focus on providing great content and on growing your listeners and monetization will take care of itself.

With that being said, we're now at the end of this book. I'm positive you've learned a lot from this in-depth look at the various aspects of podcasting and look forward to hearing from you about the launch of your new podcast! I wish you the best of luck and all the success in the world!

References

Advertise Your Podcast. (2019). Retrieved 19 November 2019, from https://overcast.fm/account/buy_ad

Booth, I. (2019). E-A-T and SEO: How to Create Content That Google Wants. Retrieved 19 November 2019, from https://moz.com/blog/google-e-a-t

Fang, W. (2019). How Do Podcasts Make Money in 2019? Here Are 8 Intriguing Ways. Retrieved 19 November 2019, from https://www.listennotes.com/podcast-academy/how-do-podcasts-make-money-in-2019-here-are-8-2/

Geoghegan, M., & Klass, D. (2008). Podcast Solutions. Berkeley, CA: Apress, Inc.

Gibson, K. (2018). America's most hated companies. Retrieved 19 November 2019, from https://www.cbsnews.com/news/americas-most-hated-companies/

Gregory, D. (2017). How to Choose the Best Media Contact Database in 2017. Retrieved 19 November 2019, from https://www.sitevisibility.co.uk/blog/2017/08/14/choose-best-media-contact-database-2017/

Maclean, M. (2017). The Best Podcast Hosting Services: Where to Host your Podcast. Retrieved 19 November 2019, from https://www.thepodcasthost.com/websites-hosting/best-podcast-hosting/

Mckay, T. (2019). Twitter CEO Jack Dorsey: I Suck and the Problem Is the Whole Site. Retrieved 19 November 2019, from https://gizmodo.com/twitter-ceo-jack-dorsey-i-suck-and-the-problem-is-the-1832578727

Mclean, M. (2018). How to Start a Podcast: Every Single Step. Retrieved 19 November 2019, from https://www.thepodcasthost.com/planning/how-to-start-a-podcast/

Moazzez, N. (2018). How To Make Money Podcasting: 18 Ways To Monetize A Podcast (2019). Retrieved 19 November 2019, from https://navidmoazzez.com/make-money-podcasting/

Stokes, C. (2019). Google's Mission Statement: The Key to Long-Term SEO Success | SEO.com. Retrieved 19 November 2019, from https://www.seo.com/blog/googles-mission-statement-key-longterm-seo-success/

Made in the USA
Coppell, TX
05 July 2022

79589682R00074